THE SUNDAY TIMES

TEASERS

BOOK 1

EDITED BY JOHN OWEN

Published in 2021 by Times Books
HarperCollins*Publishers*
Westerhill Road
Bishopbriggs
Glasgow, G64 2QT

HarperCollins*Publishers*
Macken House, 39/40 Mayor Street Upper,
Dublin 1, D01 C9W8, Ireland

www.collinsdictionary.com

© Times Newspapers Limited 2021

10 9 8 7 6 5 4 3

The Sunday Times is a registered trademark of Times Newspapers Ltd

ISBN 978-0-00-847278-8

Typeset by Davidson Publishing Solutions, Glasgow

Printed and bound in the UK using 100% Renewable
Electricity at CPI Group (UK) Ltd

Images used under license from Shutterstock.com

If you would like to comment on any aspect of this book, please contact us at
the above address or online via email: puzzles@harpercollins.co.uk

🐦 Follow us on Twitter @collinsdict
📘 Facebook.com/CollinsDictionary

This book contains FSC™ certified paper and other controlled
sources to ensure responsible forest management.

For more information visit: www.harpercollins.co.uk/green

INTRODUCTION

Welcome to this new collection of Sunday Times Teasers. Here we bring together 100 mind-bending puzzles from the series, which has been tantalising mathematically inclined readers for more than 60 years.

Each Teaser takes the form of a paragraph or two of text, sometimes accompanied by a diagram. The puzzles invariably require the application of mathematical or logical reasoning to reach the answer, and, like any true intellectual challenge, completing one offers great satisfaction to the solver.

In this collection we also provide full solutions to each of the puzzles, over and above the short answer that we give in the newspaper. These solutions don't just reveal the answer, but also show exactly how to get the answer, which I hope will provide some enlightenment to the truly puzzled.

The Teasers in this book were originally published between January 2019 and December 2020, and all have been edited by the brilliant John Owen. I am immensely grateful for the work John puts into each puzzle before it appears in the paper and the efforts he has made in producing this book.

John also contributes puzzles to the series (you can see some of his work on the pages of this collection), complementing those submitted by many other regular and occasional setters. The seemingly endless ingenuity of our setters is nothing short of astonishing, and I would like to thank them all for the entertainment they continue to provide.

All Teaser submissions are unsolicited and anyone with a good idea is welcome to send in a puzzle. If you have been inspired by the challenges in this book and would like to write a Teaser puzzle for publication in *The Sunday Times*, please email: puzzles.feedback@sunday-times.co.uk for more information.

David Parfitt
Puzzles Editor, The Times & The Sunday Times

PUZZLES

1

LONG DIVISION

Graham Smithers

I wrote down a 2-digit number and a 5-digit number and then carried out a long division.

I then erased all of the digits in the calculation, other than the 1's, and so finished up with the following:

```
           1  .   .   .
           _____
 .  .  ( .   .   .   . 1
        _____
        .   .
        _____
        .  1  .
        1  .   .
        _____
        .   .   .
        _____
           1  1  1
           1  1  1
        _____
```

If I told you how many different digits I had erased, then you should be able to work out my two original numbers.

What were my two original numbers?

2

BETTING TO WIN

Mike Fletcher

Two teams, A and B, play each other.

A bookmaker was giving odds of 8-5 on a win for team A, 6-5 on a win for team B and X-Y for a draw (odds of X-Y mean that if £Y is bet and the bet is successful then £(X + Y) is returned to the punter). I don't remember the values of X and Y, but I know that they were whole numbers less than 20.

Unusually, the bookmaker miscalculated! I found that I was able to make bets of whole numbers of pounds on all three results and guarantee a profit of precisely 1 pound.

What were the odds for a draw, and how much did I bet on that result?

3

BIG DEAL

Victor Bryant

My wife and I play bridge with Richard and Linda. The 52 cards are shared equally among the four of us – the order of cards within each player's hand is irrelevant but it does matter which player gets which cards. Recently, seated in our fixed order around the table, we were discussing the number of different combinations of cards possible and we calculated that it is more than the number of seconds since the 'big bang'!

We also play another similar game with them, using a special pack with fewer cards than in the standard pack – again with each player getting a quarter of the cards. Linda calculated the number of possible combinations of the cards for this game and she noticed that it was equal to the number of seconds in a whole number of days.

How many cards are there in our special pack?

4

WHAT DO POINTS MAKE?

Nick MacKinnon

In the Premier League table a team's points are usually roughly equal to their goals scored (Burnley were an interesting exception in 2017-18). In our football league after the four teams had played each other once, with 3 points for a win and 1 for a draw, each team's points were exactly equal to their goals scored.

A ended up with the most points, followed by B, C and D in that order. Fifteen goals had been scored in total, and all the games had different scores. One game finished 5-0, and the game BvD had fewer than three goals.

What were the results of B's three games (in the order BvA, BvC, BvD)?

5

KEEP YOUR BALANCE

Danny Roth

George and Martha have a set of a dozen balls, identical in appearance but each has been assigned a letter of the alphabet, A, B....K, L and each is made of a material of varying density so that their weights in pounds are 1, 2 ..... 11, 12 but in no particular order. They have a balance and the following weighings were conducted:

(1) A + C + I v G + J + L
(2) A + H + L v G + I + K
(3) B + I + J v C + F + G
(4) B + D + I v E + G + H
(5) D + F + L v E + G + K

On all five occasions, there was perfect balance and the total of the threesome in each was a different prime number, that in 1) being the smallest and progressing to that in 5) which was the largest.

In alphabetical order, what were the weights of each of the twelve balls?

6

HAPPY FAMILIES

Angela Newing

Oliver lives with his parents, Mike and Nellie, at number 5.
In each of numbers 1 to 4 lives a family, like his, with a mother,
a father and one child. He tries listing the families in alphabetical
order and produces a table thus:

House number	1	2	3	4	5
Father	Alan	Dave	George	John	Mike
Mother	Beth	Ellen	Helen	Kate	Nellie
Child	Carol	Freddie	Ingrid	Larry	Oliver

However, apart from his own family, there is just one father, one
mother and one child in the correct position. Neither Helen nor
Beth lives at number 3 and neither Dave nor Ingrid lives at
number 1. George's house number is one less than Larry's and
Beth's house number is one less than Carol's.

Apart from Oliver's family, who are correctly positioned?

7

INFERNAL INDICES

Stephen Hogg

$$6^{\left[6^6\right]} - \left[6^6\right]^6$$

is the size of the evil multitude in the last 'Infernal Indices' novel 'A deficit of daemons'.

'Spoiler alert' – at the end, the forces of good engage in the fewest simultaneous battles that prevent this evil horde splitting, wholly, into equal-sized armies for that number of battles. The entire evil horde is split into one army per battle, all equally populated, bar one which has a deficit of daemons, leading to discord and a telling advantage for the forces of good.

How many battles were there?

PENTAGONAL GARDENS

Nick MacKinnon

Adam and Eve have convex pentagonal gardens consisting of a square lawn and paving. Both gardens have more than one right-angled corner. All the sides of the gardens and lawns are the same whole number length in metres, but Adam's garden has a larger total area. Eve has worked out that the difference in the areas of the gardens multiplied by the sum of the paved areas (both in square metres) is a five-digit number with five different digits.

What is that number?

55 DIVISIONS

Tom Wills-Sandford

To mark her 55th birthday, Martha, a school teacher, gave each of her nine pupils a sheet of paper with a single different digit from 1 to 9 written on it.

They stood at the front of the classroom in a row and the 9-digit number on display was divisible by 55. Martha then asked the first 3 in the row (from the left) to sit down. The remaining 6-digit number was also divisible by 55. The next 3 then sat down and the remaining 3-digit number was also divisible by 55.

The 9-digit number was the smallest possible. What was it?

A HARDY ANNUAL

Victor Bryant

Yesterday (23/3/2019) was my grandson's birthday and we continued a family tradition. I asked him to use any eight different non-zero digits (once each) to form a set of numbers that added to 2019. Last year I asked the equivalent question with a sum of 2018, and I have done this each year for over ten years. Only on one occasion has he been unable to complete the task.

In this year's answer his set of numbers included a 3-figure prime that had also featured in last year's numbers.

(a) In which year was he unable to complete the task?

(b) What was the 3-figure prime that featured in this year's answer?

11

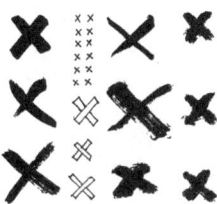

MYSTERY NUMBERS

Graham Smithers

I wrote down a 2-digit number and a 5-digit number and then carried out a long division.

I then erased all of the digits in the calculation, other than one of them, which I have indicated by #.

This gave me the following:

```
              .     .      .       .
    .   #(   #    #    .    .    #
            .    #
        .    .    .
            .     .
        .    .    .
        .   #    .
              .     #
              .     #
```

What were my two original numbers?

12

TEN DIGITS

Andrew Skidmore

Without repeating a digit I have written down three numbers, all greater than one. Each number contains a different number of digits. If I also write down the product of all three numbers, then the total number of digits I have used is ten. The product used only two different digits, each twice, neither of which appears in the three original numbers.

What is the product?

13

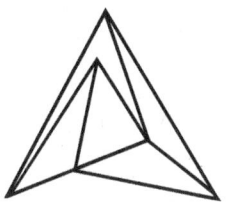

IMPRISMED

Stephen Hogg

A right regular prism has two ends with identical faces, joined by oblong rectangular faces. I have eight of them, with regular convex polygonal end-faces of 3, 4, 5, 6, 7, 8, 9 and 10 sides (triangle, square and so on). They sit on my flat desk (on oblong faces), and each prism has the same height.

I chose three prisms at random, and was able to slide them into contact, broadside, in such a way that the middle one overhung both others (and could be lifted without disturbing them). Also, I was able to slide one outer prism to the other side, and the new 'middle' prism was overhung by both others (and so vertically 'imprisoned' by them).

I was able to do all this again with three randomly chosen remaining prisms.

Give the prior chance of this double selection (as a fraction in lowest terms)

14

SILLY SLIP

Victor Bryant

Please help me find my silly slip. I correctly added a five-figure number to a four-figure number to give a six-figure total. Then I tried to substitute letters for digits systematically and I ended up with

SILLY + SLIP = PLEASE

However, in one of these letters I have made a silly slip, so please find it and then work out what the correct sum was.

What was the correct six-figure numerical answer?

15

MARBLE TOWER

Andrew Skidmore

Liam has a number of bags of marbles; each bag contains the same number (more than 1) of equal-size marbles.

He is building a tetrahedron with the marbles, starting with a layer that fits snugly in a snooker triangle. Each subsequent triangular layer has one fewer marble along each edge. With just one full bag left he had completed a whole number of layers; the number of marbles along the edge of the triangle in the last completed layer was equal to the number of completed layers. The last bag had enough marbles to just complete the next layer.

How many bags of marbles did Liam have?

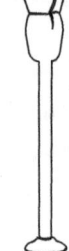

LOVELY METER, RITA MADE!

Stephen Hogg

Revisiting the old curiosity shop, I bought an unusual moving-iron ammeter (made by Rho Iota Tau Associates). On the non-linear scale '9' was the last possible 'whole amperes' scale graduation marked before the 'full scale deflection' end stop (which was a half turn of the pointer from zero).

The booklet gave the pointer swing from zero (in degrees) equal to the current (in amperes) raised to a fixed single-figure positive whole number power, then divided by a positive whole number constant. Curiously, the angle between 'exact' pointer swings, calculated using this formula, for two single-figure 'whole amperes' values was exactly a right angle.

What, in amperes, were these two currents (lower first) and to what power were they raised?

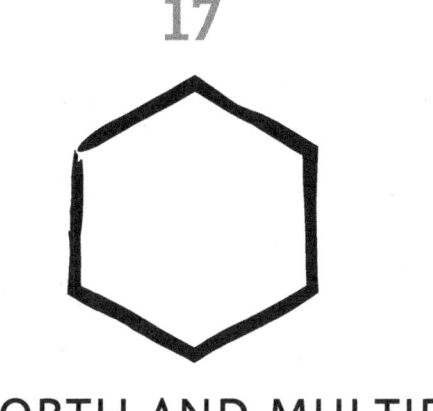

GO FORTH AND MULTIPLY

Nick MacKinnon

Adam and Eve have convex hexagonal gardens whose twelve sides are all the same whole number length in yards. Both gardens have at least two right-angled corners and the maximum possible area this allows. Each garden has a path from corner to corner down an axis of symmetry. Adam multiplies the sum of the path lengths by the difference of the path lengths (both in yards) and Eve squares Adam's answer, getting a perfect fifth power with no repeated digit.

What was Eve's answer?

A NICE LITTLE EARNER

Graham Smithers

The "value" of a number is found by subtracting its first digit from the last. For example, 6, 72, 88 and 164 have values 0, - 5, 0 and 3 respectively.

Raising funds for a local charity, I placed some raffle tickets numbered from 1 up to a certain 3-digit number, in a box. Participants then selected a ticket at random.

If the value of their number was positive, they won that amount in £; if the value was negative, they contributed that amount in £. Otherwise no money changed hands.

All the tickets having been used, the total amount raised in £ was a rearrangement of the digits in that number of tickets.

How much was raised?

19

BEYOND THE FIELDS WE KNOW

Stephen Hogg

A field named 'Dunsany levels' has four unequal straight sides, two of which are parallel. Anne – with her dog, Newton – walked from one corner of the field straight towards her opposite corner. Leon did the same from an adjacent corner along his diagonal. Yards apart, they each rested, halfway along their paths, where Leon, Anne and a signpost in the field were perfectly aligned. Straight fences from each corner converged at the signpost, making four unequal-area enclosures.

Newton made a beeline for the signpost, on which the whole-number area of the field, in acres, was scratched out. Clockwise, the enclosures were named: 'Plunkett's bawn', 'Three-acre meadow', 'Drax sward' and 'Elfland lea'. Anne knew that 'Three-acre meadow' was literally true and that 'Elfland lea' was smaller by less than an acre.

What was the area of 'Dunsany levels' in acres?

20

SUDOKU CROSSWORD

Danny Roth

George and Martha are doing a mathematical crossword. There is a 3 x 3 grid with the numbers 1 to 9 inclusive appearing once each. The clues are as follows:

Across:	top line:	a prime number
	middle line:	a prime number
	bottom line:	a prime number

Down:	left column:	a perfect square
	middle column:	a perfect square
	right column:	a prime number

Although you do not need to know this, one of the diagonal three-digit numbers is also prime.

What is the sum of the three "across" numbers?

THE MAGNIFICENT SEVEN

Graham Smithers

After a day's filming, a group of those involved in the film's production went for a gallop.

They split into threes, with a samurai leading each grouping.

The seven samurai were:
BRONSON, BRYNNER, BUCHHOLZ, COBURN, DEXTER, MCQUEEN and VAUGHN

The others involved in the gallop were:
ALANIZ, ALONZO, AVERY, BISSELL, BRAVO, DE HOYOS, HERN, LUCERO, NAVARRO, RUSKIN, RUSSELL, SUAREZ, VACIO and WALLACH

For each grouping, any two names from the three had exactly 2 letters in common (e.g. BRYNNER and BRAVO have B and R in common).

If I told you who accompanied BRONSON, you should be able to tell me who accompanied (a) MCQUEEN and (b) DEXTER

22

BAG OF SWEETS!

Angela Newing

I recently bought a number of equally-priced bags of sweets for a bargain price, spending more than 50p in total. If they had been priced at 9p less per bag, I could have had 2 bags more for the same sum of money. In addition, if each had cost 12p less than I paid, then I could also have had an exact number of bags for the same sum of money.

How much did I spend in total on the sweets?

23

ALAN AND CAT

John Owen

Alan and Cat live in a city that has a regular square grid of narrow roads. Avenues run west/east, with 1st Avenue being the furthest south, while Streets run south/north with 1st Street being the furthest west.

Cat lives at the intersection of 1st Street and 1st Avenue, while Alan lives at an intersection due northeast from Cat. On 1 January 2018, Cat walked to Alan's house using one of the shortest possible routes (returning home the same way), and has done the same every day since. At first, she walked a different route every day and deliberately never reached an intersection where the Street number is less then the Avenue number. However, one day in 2019 she found that she could not do the same, and repeated a route.

What was the date then?

24

BOOK RECEIPTS

Howard Williams

My wife recently purchased two books from her local bookshop. She showed me the receipt, which showed the cost of each book and the three-figure total cost. I noticed that all of the digits from 1 to 9 had been printed. Coincidentally, exactly the same happened to me when buying two different books, but my more expensive book cost more than hers. In fact, it would not have been possible for that book to have cost more.

How much did I pay for the more expensive book?

25

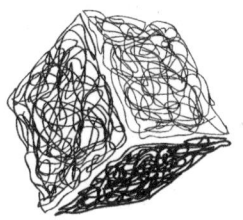

A RESULT

Graham Smithers

For any number, I square the digits and then add the resulting numbers. If necessary, I keep repeating the process until I end up with a single digit, called the result.

For example: 142 gives $1 + 16 + 4 = 21$ which then gives $4 + 1 = 5$, the result.

I have written down a two-digit number. If I tell you one of the digits [the key digit], you should be able to work out the result.

I then use a 3rd digit to get a three-digit number. The result of that number happens to be the key digit.

In increasing order, what are the three digits?

'BINGO A GO-GO' LINGO A NO-GO

Stephen Hogg

My rest home's Bingo set uses numbers 1 to 99. To win, nine numbers on your game card must be called. Our caller, not knowing 'bingo-lingo', says 'Number 1, total factors 1', 'Number 11, total factors 2' and 'Number 30, total factors 8', etc.

Yesterday, in one game, my hearing aid howled whenever a call started. I missed each number, but heard each 'total factors' value. Fortunately, after just nine calls I shouted 'HOUSE!' certain that I'd won.

I told my daughter how many different 'factor' values I'd heard, but didn't say what any of the values were. Knowing that I had won after nine calls, she could then be sure about some (fewer than nine) of my winning numbers.

Give, in ascending order, the numbers that she could be sure about?

KNOCK, KNOCK!

Victor Bryant

Last year a two-figure number of teams entered our football competition. With that number, it was impossible to have a straightforward knockout competition (where half the teams are knocked out in each round), so we divided the teams into equal-sized groups, each pair of teams within a group played each other once, and the overall winner from each group went forward to the second stage consisting of a knockout competition. Unfortunately our local team was knocked out in the quarter-finals of that stage.

This year a higher number of teams entered (but not twice as many). Again a straightforward knockout competition was impossible so we copied last year's model but with smaller groups and more of them. By coincidence the total number of games played this year equalled the number played last year.

How many teams entered last year and how many this year?

28

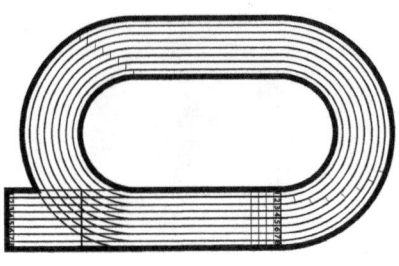

ON TRACK

Howard Williams

Sarah and Jenny are runners who train together on a circular track, and Sarah can run up to 20% faster than Jenny. They both run at a constant speed, with Sarah running an exact percentage faster than Jenny. To reduce competition they start at the same point, but run round the track in different directions, with Sarah running clockwise.

On one day they passed each other for the seventh time at a point which is an exact number of degrees clockwise from the start. Sarah immediately changed her pace, again an exact percentage faster then Jenny. After a few passes both runners reached the exit, at a point on the track an exact number of degrees clockwise from the start, at the same time.

How fast, relative to Jenny, did Sarah run the final section?

ODDS AND EVENS

Angela Newing

I have done a 'long multiplication', which is reproduced below.
[If the multiplication was ABC x DE, then the third line shows the
result of ABC x E and the fourth line shows the result of ABC x D].
However instead of writing the actual digits involved, I have
written 'o' where there is an odd digit and 'e' where there is an
even digit.

		o	e	e
x			e	e
e	o	e	e	
e	o	e	.	
o	o	e	e	

What is the result of the multiplication?

30

GARDENING DIVISION

Danny Roth

George and Martha's garden is a perfect square of length (whole number of metres) a two-digit number ab. The area is a three-digit number cde. In the middle, they have planted a square flowerbed of a length which is a single-digit number f and area a two-digit number gh.

They have called in a gardener, who works for a single-digit i hours. He works for a whole number of minutes on the flowerbed and the remainder on the surrounding lawn. Each square metre of the flowerbed requires n (a single digit) times the time spent on each square metre of the surrounding lawn. I have mentioned nine letters, a-i inclusive, and each stands for a different positive digit.

How many minutes does the gardener work on the lawn?

31

SLIDE RULES

Stephen Hogg

Using her ordinary 15cm ruler and Zak's left-handed version (numbers 1 to 15 reading right to left) Kaz could display various fractions. For instance, putting 5 on one ruler above 1 on the other ruler, the following set of fractions would be displayed: 5/1, 4/2, 3/3, 2/4 and 1/5. Zak listed the fifteen sets starting from '1 above 1' up to '15 above 1'.

Kaz chose some fractions with values less than one from Zak's sets (using just the numerals 1 to 9, each once only in her selection). Of these, two were in simplest form, one of which had consecutive numerator and denominator. Zak correctly totalled Kaz's selection, giving the answer as a fraction in simplest form. Curiously, the answer's numerator and denominator were both palindromic.

Give Zak's answer.

PUZZLING POWERS

Andrew Skidmore

I came across a most remarkable number recently and wrote it down. All its digits were different and it was divisible by 11. In itself, that wasn't particularly interesting, but I wrote down the number of digits in the number and then wrote down the sum of the digits in the number.

I therefore had three numbers written down. What surprised me was that of the three numbers, one was a square and two were cubes.

What is the remarkable number?

33

SIX SISTERS ON THE SKI LIFT

Bernardo Recaman

The sum of the ages of six sisters known to me is 92.
Though there is no single whole number greater than 1 that
simultaneously divides the ages of any three of them, I did
notice this morning, while they lined up for the ski lift, that the
ages of any two of them standing one behind the other, had a
common divisor greater than 1.

In increasing order, how old are the six sisters?

34

A RELAXING DAY

Danny Roth

George and Martha have a digital clock, which displays time with six digits on the 24-hour system, i.e. hh:mm:ss.

One afternoon, George looked at the clock and saw a six-digit display involving six different positive digits. He dozed off immediately, and when he awoke in the evening he saw another display of six digits, again all positive and different. He dozed off immediately and later on (before midnight) he awoke, having slept for exactly 23 minutes longer than the previous time. At that time, he saw a third display, yet again comprising six different positive digits. He thus had seen eighteen digits and the nine positive digits had each appeared exactly twice.

At what time did George wake up after his first sleep?

35

SOMETHING IN COMMON

Victor Bryant

I have written down four different numbers. The third number is the highest common factor of the first two (i.e. it is the largest number that divides exactly into both of them). The fourth number is the lowest common multiple of the first two (i.e. it is the smallest number that both of them divide exactly into).

I can consistently replace digits by letters in my numbers so that the highest common factor is HCF and the lowest common multiple is LCM.

What are the first two numbers?

36

FLOCKBUSTER

Stephen Hogg

Wu, Xi, Yo and Ze had different two-figure numbers of sheep and kept them in a walled field divided by fences into a fold each. Maths whizz, Wu, with the largest flock, noticed that together her flock and Ze's equalled Xi's and Yo's combined; and that, as a fraction, the ratio of Yo's flock to Xi's had consecutive upper and lower numbers (e.g. 3/4), whereas her flock to Xi's ratio had those numbers swapped over (e.g. 4/3).

Overnight, storm-damaged fences led to the same number of sheep in each fold. Wu's old maths' teacher, told just this number and the above relationships, couldn't be certain how many sheep Wu owned (which would have been true, also, if he'd been told either fraction instead).

How many sheep did Wu own?

37

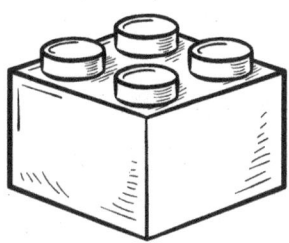

HOLLOW CUBE LAND

Bill Kinally

I have a large box of toy building bricks. The bricks are all cubes (the same size), and can be pushed together then dismantled.

I decided to build the largest cube possible by leaving out all the interior bricks. When my hollow cube was finished I had two bricks left over. I put all the bricks back in the box and gave it to my two children. Each in turn was able to use every brick in the box to construct two hollow cubes, again with all interior bricks removed. Their cubes were all different sizes.

I told them this would not have been possible had the box contained any fewer bricks.

How many bricks were in the box?

PIECE OF CAKE

Stephen Hogg

Using her special recipe, so that each cubic inch of baked cake weighed one ounce, mam made the cake for my eighth birthday party. It had a regular octagonal flat top and base, and equal square vertical faces, several inches high exactly.

Not all (but a majority) of the dozen invited pals came to the party. We each had an equal portion of cake (the largest whole number of ounces possible from it, a two-figure number). Mam had the leftover cake. Curiously, if one more or one fewer pal had turned up and our portions had been worked out in the same way, mam's leftover would have been the same in each case, but less than she actually got.

How large, in ounces, was my portion?

39

ENJOYING RETIREMENT

Danny Roth

George and Martha have worked on separate departments of a company which has four-digit telephone extensions. George looked at his extension and it was abcd. Martha's (larger) also had a, b and c as her first three digits but not necessarily in that order. Her last digit was e. They added up their two four-digit numbers and found that the least significant digit was f. They then looked at the difference and that was a four-digit number of which the least significant digit was g. They then looked at the product and the least significant digit was h. They then looked at the average of the extensions (a whole number); in it the first two digits were equal, the last two digits were also equal, and the least significant digit was i. I have thus mentioned nine digits, all positive and unequal.

What was Martha's extension?

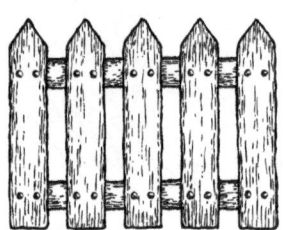

NORFOLK FLATS

Andrew Skidmore

Sam has purchased Norfolk Flats; an area of farmland (less than 100 hectares) bordered by six straight fences of the same length. He intends to farm an area which is an equilateral triangle with corners that are the midpoints of three of the existing boundaries. This creates three more distinct areas (one for each of his sons); these areas are identical in shape and size and have two sides that are parallel.

Sam measured the area (in square metres) which each son will farm and also his own area. One of the numbers is a square and the other a cube. If I told you which was which, you should be able to work out the area of Norfolk Flats.

What (in sq metres) is that area?

41

MISCHIEVOUS SAM

Graham Smithers

I set Sam a question, the answer to which was a 3-digit number, with the digits increasing by 1 from first to last (e.g. 789)

Sam eventually produced a 3-digit answer, but only two of his digits were correct and in the correct position. The third digit was wrong.

Investigating further I found that Sam had the correct answer but, for devilment, decided to change it into a different (single-digit) base.

If I were to tell you which of his 3 digits was the wrong one, you should be able to tell me:

(a) the correct answer and (b) the base used by Sam

42

EGYPTIAN WEIGHTS
AND MEASURES

Susan Bricket and John Owen

We were wondering why ancient Egyptians chose to represent arbitrary fractions as sums of distinct *unit* fractions of the form 1/n (thus 5/7 = 1/2+1/5+1/70). One of us recalled long ago watching our greengrocer use four brass weights of 1/2, 1/4, 1/8, 1/16 lb to weigh any number of ounces up to 15 by stacking some of them on one side of her balancing scales. We wanted to make a metric equivalent, a set of distinct weights of unit fractions of a kilo, each weighing a whole number of grams, to weigh in 10g steps up to 990g.

Naturally, we wanted to use as little brass as possible, but we found that there were several possible such sets. Of these, we chose the set containing the fewest weights.

List, in increasing order, the weights in our set.

FAULTY PEDOMETER

Howard Williams

Judith is a keen walker who uses a five-digit pedometer to record her number of steps. Her pedometer is inaccurate as some of the counters consistently move on to 0 early by missing out one or more digits. For instance, one of them might roll over from 7 to 0 every time instead of from 7 to 8, missing out digits 8 and 9. She is, however, well aware of this and can work out the correct number of steps.

After walking her usual distance, the pedometer shows 37225 steps but she knows that the true number is 32% less than this. A second distance she walks requires a 30% reduction in the number displayed to give the true number of steps.

How many steps is the second distance?

44

ANTIQUE TEAPOT

Graham Smithers

My local antiques dealer marks each item with a coded price tag in which different digits represent different letters. This enables him to tell the whole number of pounds he paid for the item.
I bought a teapot from him tagged MOE.

Inside the teapot was a scrap of paper, which I used to work out his code. The letters AMOUNT I SPENT had been rearranged to make the following multiplication sum:

```
        M  A  N
           P  I  N
     T  U  O  N
   A  A  S  A
T  M  E  P
T  O  O  T  I  N
```

[MAN x N = TUON, etc]

How much did he pay for the teapot?

45

MATHS FOR DESSERT

Danny Roth

George and Martha have their five daughters round the dinner table. After the meal, they had ten cards numbered 0 to 9 inclusive and randomly handed two to each daughter. Each was invited to form a two-digit number. The daughter drawing 0 obviously had no choice and had to announce a multiple of ten.

However, the others each had the choice of two options. For example if 3 and 7 were present, either 37 or 73 would be permissible. George added up the five two-digit numbers (exactly one being divisible by 9) and Martha noticed that three of the individual numbers divided exactly into that total.

What was the total of the remaining two numbers?

46

SHUFFLE THE CARDS

Victor Bryant

In the classroom I have a box containing ten cards each with a different digit on it. I asked a child to choose three cards and to pin them up to make a multiplication sum of a one-figure number times a two-figure number. Then I asked the class to do the calculation.

During the exercise the cards fell on the floor and a child pinned them up again, but in a different order. Luckily the new multiplication sum gave the same answer as the first and I was able to display the answer using three of the remaining cards from my box.

What was the displayed answer?

WHAT'S MY (LAND) LINE?

Angela Newing

My telephone has the usual keypad:

```
1 2 3
4 5 6
7 8 9
  0
```

My 11-digit telephone number starts with 01 and ends in 0.
All digits from 2 to 9 are used exactly once in between, and each
pair of adjacent digits in the phone number appear in a different
row and column of the keypad array.

The 4th and 5th digits are consecutive as are the 9th and 10th
and the 8th digit is higher than the 9th.

What is my number?

THE PRHYMES' NUMBERS

Stephen Hogg

The Prhymes' triple live album 'Deified' has hidden numerical 'tricks' in the cover notes. Track 1, with the shortest duration, is a one-and-a-half minute introduction of the band, shown as '1. Zak, Bob, Kaz 1:30' in the cover notes. The other nineteen tracks each have different durations under ten minutes and are listed similarly with durations in 'm:ss' format.

For each of tracks 2 to 20, thinking of its 'm:ss' duration as a three-figure whole number (ignoring the colon), the track number and duration value each have the same number of factors (counting all factors, not just prime ones). Curiously, the most possible are palindromic; and the most possible are even.

What is the total album duration (given as m:ss)?

49

TABLE MATS

Graham Smithers

Beth and Sam were using computers to design two table mats shaped as regular polygons.

The interior angles of Beth's polygon were measured in degrees; Sam's were measured in grads. [A complete turn is equivalent to 360 degrees or 400 grads.]

On completion they discovered that all of the interior angles in the 2 polygons had the same numerical whole number value.

If I told you the last digit of that number, you should be able to work out the number of edges in (a) Beth's table mat and (b) Sam's table mat.

50

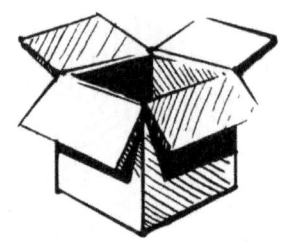

OPEN THE BOX

John Owen

Game show contestants are shown a row of boxes, each containing a different sum of money, increasing in regular amounts (e.g. £1100, £1200, £1300, ...), but they don't know the smallest amount or the order. They open one box then take the money, or decline that and open another box (giving them the same choices, apart from opening the last box when they have to take the money).

Alf always opens boxes until he finds, if possible, a sum of money larger than the first amount. Bert's strategy is similar, except he opens boxes until he finds, if possible, a sum of money larger than both of the first two amounts. Remarkably, they can both expect to win exactly the same amount on average.

How many boxes are there in the game?

POT LUCK

Andrew Skidmore

Liam had a bag of snooker balls containing 6 'colours' (not red) and up to 15 red balls. He drew out balls at random, the first being a red. Without replacing this he drew another ball; it was a 'colour'. He replaced this and drew another ball. This was a red (not replaced), and he was able to follow this by drawing a 'colour'. The probability of achieving a red/colour/red/colour sequence was one in a certain whole number.

After replacing all the balls, Liam was able to 'pot' all the balls. This involved 'potting' (i.e. drawing) red/colour/red/colour...red/colour (always replacing the colours but not the reds), then 'potting' the six colours (not replaced) in their correct sequence. Strangely, the probability of doing this was also one in a whole number.

What are the two whole numbers?

52

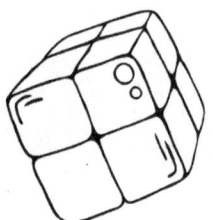

SQUARES AND CUBES

Howard Williams

Jenny is pleased that she has found two whole numbers with a remarkable property. One of them is a single digit greater than zero while the other one has two digits. The remarkable thing is that the difference of their squares is a perfect cube and the difference of their cubes is a perfect square.

Her sister Sarah is not impressed, however. She has found two three-digit numbers for which the difference of their squares is also a perfect cube and the difference of their cubes is also a perfect square.

In ascending order, what are the four numbers?

53

SUPER STREET

Dany Roth

George and Martha and their five daughters with their families have moved into six houses in Super Street. I define a 'super prime number' as a prime number that has two or more digits whose sum is a prime number (e.g. 29 and 101). Similarly for 'super squares' (e.g. 36) and 'super cubes'. Houses in Super Street are numbered with the lowest 31 super numbers of the above types.

The elderly couple live in the highest-numbered house on the street. They noticed that the last digits of their daughters' houses were five consecutive digits and the sum of their five house numbers was a perfect square. Furthermore, the ordinal positions (lowest-numbered house is 1 and so on) of all but one of the houses were prime.

Which five houses did the daughters occupy?

CRYSTAL CLEAVAGE CALAMITY!

Stephen Hogg

My 'mystic' crystal stood on its triangular end face, with every side of the triangle being less than 10cm. It had three vertical sides and an identical triangular top end face, with the height being under 50cm. Unfortunately, during dusting, it fell and broke in two. The break was a perfect cleavage along an oblique plane that didn't encroach on the end faces, so I then had two, pointed, pentahedral crystals.

Later, I found that the nine edge lengths, in cm, of one of these were all different whole numbers – the majority prime – and their total was also a prime number. Curiously, the total of the nine, whole-number, edge lengths, in cm, of the other piece was also a prime number.

What was the total for this latter piece?

55

BAKER'S WEIGHTS

Peter Good

A baker's apprentice was given a 1kg bag of flour, scales and weights, each engraved with a different whole number of grams. He was told to separately weigh out portions of flour weighing 1g, 2g, 3g, and so on up to a certain weight, by combining weights on one side of the scales. He realised that he couldn't do this if there were fewer weights, and the largest weight was the maximum possible for this number of weights, so he was surprised to find after one of these weighings that the whole bag had been weighed out exactly. Upon investigation, he discovered that some of the weights weighed 1g more than their engraved weight. If I told you how many of the weights were too heavy, you would be able to work out what they all were.

List all of the true weights (in ascending order)?

56

CONSECUTIVE SUMS

Bill Kinally

Amelia noticed that 15 is equal to 1+2+3+4+5 or 4+5+6 or 7+8, so there are three possible ways that it can be expressed as the sum of consecutive whole numbers. She then told Ben that she had found a three-digit number which can be expressed as the sum of consecutive whole numbers in just two different ways. "That's interesting", said Ben. "I've done the same, but my number is one more than yours".

What is Ben's number?

PUB CELEBRATION

Howard Williams

While celebrating with friends in the pub, it was my round, and I worked out the number of ways that I could get everybody else's drink wrong. I could remember the drinks (and they all wanted different drinks), but not who wanted which drink.

More friends arrived (there were fewer than twenty of us in total), and everyone still wanted different drinks. Again assuming that I could remember the drinks but not who wanted which drink, I worked out the number of ways that I could get everybody else's drink wrong, and found that the number was 206 times the first number. Fortunately, it was no longer my round!

What was the second number?

PIECE IT TOGETHER

Victor Bryant

I have some jigsaw-type pieces each consisting of one, two, three or four 1cm-by-1cm squares joined together without overlapping. The pieces are black on one side and white on the other, and they are all different. I have used all my pieces to simultaneously make some different-sized white squares in jigsaw fashion, with each square using more than one piece. Even if you knew what all my pieces were like, you would not be able to determine the sizes of all of my squares.

How many pieces do I have?

CEMETERY LOTTERY SYMMETRY

Stephen Hogg

Our local cemetery conservation lottery tickets have four numbers for a line. Using eight different numbers, my two lines have several symmetries. For each line: just one number is odd; there is one number from each of the ranges 1-9, 10-19, 20-29 and 30-39, in that order; the sum of the four numbers equals that sum for the other line; excluding 1 and the numbers themselves, the 1st and 2nd numbers share just one factor – as do the 2nd and 3rd (a different factor) and the 3rd and 4th (another different factor) and, finally, the 4th and 1st.

Printed one line directly above the other, my top line includes the largest of the eight numbers.

What is the bottom line?

60

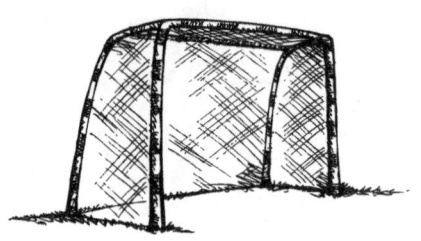

FOOTBALL LEAGUE

Andrew Skidmore

In our league, each team is awarded three points for a win and one for a draw. Teams play each other twice per season. Comparison of the results of the two Divisions at the end of last season showed that

(a) Division II held one more team than Division I.

(b) The same total number of points was awarded in each division.

(c) For the division with the larger number of drawn matches, that number was equal to the number of matches not drawn in the other division.

(d) The number of drawn matches in one division was a multiple of the (3-digit) number of drawn matches in the other.

How many teams were in Division II?

TRIANGULAR CARD TOWER

Howard Williams

Robbie leans two very thin playing cards together, then another two together, placing an identical card across the top forming a platform. He proceeds sideways, each time adding two further leaning cards and a platform card, to form one layer. He forms a second layer in the same way, but with one fewer platform card, and continues upwards with further layers to build a roughly triangular tower.

For the bottom layers, he uses a whole number of 53-card packs of large cards (integer length above 70mm), the number of packs equalling the number of bottom layers. He then uses small cards (75% length and width) to complete the tower, which is 1428mm high. The distance between the bases of two leaning cards is always 0.56 of the length of each card.

Robbie would like to extend the tower sideways and upwards to the next possible integer height, still using large cards only for the bottom layers.

How many extra cards would be needed in total?

THE THREE THOUSANDTH

Richard England

```
      T H R E E
+     T H O U S
+     A N D T H
  T E A S E R
```

In this addition digits have been consistently replaced by letters, different letters representing different digits. But instead of an addition in base 10 in which the letters represent the digits 0 to 9 this is an addition in base 11, using X for the extra digit, in which the letters represent the digits 1 to X, with 0 unused.

Please submit the number (still in base 11) represented by TEASER.

TETRAGONAL TOY TILES

Stephen Hogg

Thirteen toy tiles comprised a square, rectangles, rhombuses (diamonds on a playing card are rhombuses) and kites (as shown in the diagram). All of each different type were identical. A rhombus's longer diagonal was a whole number of inches (equalling all diagonals of all other types). Its shorter diagonal was half this. Also, one side of a rectangle was slightly over one inch.

A pattern I made, using every tile laid flat, had all the symmetries of a square. After laying the first tile, each subsequent tile touched at least one other previously placed tile. Ultimately, any contact points were only where a vertex of a tile touched a vertex of just one other tile; only rhombuses touched every other tile type.

What, in inches, was a square's diagonal?

64

SHORT-CUT

Victor Bryant

To demonstrate a bit of geometry and trigonometry to my grandson, I took a rectangular piece of paper whose shorter sides were 24 cm in length. With one straight fold I brought one corner of the rectangle to the midpoint of the opposite longer side. Then I cut the paper along the fold, creating a triangle and another piece. I then demonstrated to my grandson that this other piece had double the area of the triangle.

How long was the cut?

65

ALL THAT GLITTERS

Nick MacKinnon

My aunt has a collection of sovereigns, and she set me a challenge. "You can have the coins if you can work out their years, which (in increasing order) are equally spaced and all in the twentieth century. The number of coins is an odd prime. The highest common factor of each pair of years is an odd prime. The sum of the number of factors of each of the years (including 1 and the year itself) is an odd prime." I worked out the years, though the gift was much less valuable than I'd hoped.

What were the years?

66

GOING UP

John Owen

In our football league, the teams all play each other once, with three points for a win and one for a draw. At the end of the season, the two teams with most points are promoted, goal difference being used to separate teams with the same number of points.

Last season's climax was exciting. With just two games left for each team, there were several teams tied at the top of the league with the same number of points. One of them, but only one, could be certain of promotion if it won its two games. If there had been any more teams on the same number of points, then none could have guaranteed promotion with two wins.

How many teams were tied at the top of the league, and how many of the remaining matches involved any of those teams?

67

TUBULAR BALES

Stephen Hogg

Ten equal-length, rigid tubes, each a different prime-valued external radius from 11mm to 43mm, were baled, broadside, by placing the 43mm and 11mm tube together and the third tube, not the largest remaining, touching both of these. Each subsequent tube touched the previous tube placed and the 43mm tube. A sub-millimetre gap between the final tube placed and the 11mm tube, made a near perfect fit.

The radius sum of the first three tubes placed against the 43mm tube was a multiple of one of the summed radii. Curiously, that statement remains true when each of 'four', 'five', 'seven' and 'eight' replaces 'three'. For 'two' and 'six' tubes placed their radius sum was a multiple of an as yet unplaced tube's radius.

What radius tube, in mm, was placed last?

68

RAFFLE TICKETS

Andrew Skidmore

At our local bridge club dinner we were each given a raffle ticket. The tickets were numbered from 1 to 80. There were six people on our table and all our numbers were either prime or could be expressed as the product of non-repeating primes (e.g. 18 = 2x3x3 is not allowed). In writing down the six numbers you would use each of the digits 0 to 9 once only. If I told you the sum of the six numbers (a perfect power) you should be able to identify the numbers.

List the numbers (in ascending order).

69

PAVING STONES

Howard Williams

James has decided to lay square block paving stones on his rectangular patio. He has calculated that starting from the outside and working towards the middle that he can lay a recurring concentric pattern of four bands of red stones, then three bands of grey stones, followed by a single yellow stone band. By repeating this pattern and working towards the centre he is able to finish in the middle with a single line of yellow stones to complete the patio.

He requires 402 stones to complete the first outermost band. He also calculates that he will require exactly 5 times the number of red stones as he does yellow stones.

How many red stones does he require?

70

THREE-FRUIT FRACTIONS

Stephen Hogg

The owner of the old curiosity shop repaired an antique mechanical fruit machine having three wheels of identical size and format. Afterwards each wheel was independently fair, just as when new. Each wheel's rim had several equal-sized zones, each making a two-figure whole number of degrees angle around the rim. Each wheel had just one zone showing a cherry, with other fruits displayed having each a different single-figure number (other than one) of zone repetitions.

Inside the machine were printed all the fair chances (as fractions) of getting three of the same fruit symbol in one go. Each of these fractions had a top number equal to 1 and, of their bottom numbers, more than one was odd.

What was the bottom number of the chance for three cherries?

71

HEAD COUNT

Victor Bryant

My grandson and I play a simple coin game. In the first round we toss seven coins and I predict how many 'heads' there will be whilst my grandson predicts the number of 'tails'. After the tossing I score a point for each head plus a bonus of ten if my prediction was correct – my grandson scores likewise for the tails. We then repeat this with six coins, then five, and so on down to a single coin. After each round we keep cumulative totals of our scores.

In one game, for over half of the pairs of cumulative scores, my grandson's total was like mine but with the digits in reverse order. In fact he was ahead throughout and at one stage his cumulative total was double mine – and he had an even bigger numerical lead than that on just one earlier occasion and one later occasion.

List the number of heads in each of the seven rounds.

PUTTING GAME

John Owen

A putting game has a board with eight rectangular holes, like the example (not to scale), but with the holes in a different order.

6	3	8 7	2	5	1	4

If you hit your ball (diameter 4cm) through a hole without touching the sides, you score the number of points above that hole. The sum of score and width (in cm) for each hole is 15; there are 2cm gaps between holes.

I know that if I aim at a point on the board, then the ball's centre will arrive at the board within 12cm of my point of aim, and is equally likely to arrive at any point in that range. Therefore, I aim at the one point that maximises my long-term average score. This point is the centre of a hole and my average score is a whole number.

(a) Which hole do I aim at?
(b) Which two holes are either side of it?

73

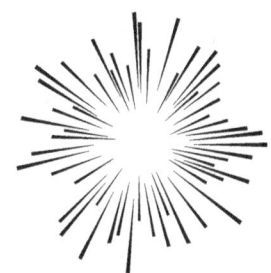

OPTICAL ILLUSION

Danny Roth

George and Martha are studying optics. If you place an object a specific distance from a lens, an image will appear at a distance from that lens according the following formula:

The reciprocal of the object distance plus the reciprocal of the image distance is equal to the reciprocal of the focal length of the lens.

The object distance was a two-digit whole number of cm (ab). The image distance and the focal length of the lens were also two-digit whole numbers (cd and ef respectively). The six digits were all different and non-zero. Also, the object distance and the focal length were of the same parity and b was an exact multiple of d. Martha pointed out that the sum of those three two-digit numbers was a prime number.

What was that prime number?

NUMBER BLIND RAGE

Stephen Hogg

After school, angry at getting '50 lines', I kicked my satchel around. Impacts made my 11-digit calculator switch on. An 11-digit number was also entered and the display was damaged. Strangely, I found 'dYSCALCULIA' displayed and saved this to memory (as shown).

After various tests I confirmed that all arithmetic operations were correct and the decimal point would appear correctly if needed. No segments were permanently 'on', two digits were undamaged, and for the other digits, overall, several segments were permanently 'off'. Results always appear at the right-hand side of the display.

Retrieving 'dYSCALCULIA', I divided it by 9, then the result by 8, then that result by 7, then that result by 6. No decimal point appeared and the last result had three digits appearing as numerals.

What number was 'dYSCALCULIA'?

75

ARIAN PEN-BLWYDD

Howard Williams

When I thought that my daughter was old enough to be responsible with money I gave her on her next, and all subsequent birthdays, cash amounts (in pounds) which were equal to her birthday age squared.

On her last birthday her age was twice the number of years for which she received no such presents. I calculated at this birthday that if I had made these gifts on all of her birthdays then she would have received 15% more than she had actually received. I then decided that I would stop making the payments after her birthday when she would have received only 7.5% more if the payments had been made on all of her birthdays.

What was the amount of the final birthday payment?

76

FAMILY BUSINESS

Danny Roth

George and Martha run a company with their five daughters. The telephone extensions all have four positive unequal digits and strangely only four digits appear in the seven extensions:

Andrea	abcd
Bertha	acdb
Caroline	bacd
Dorothy	dabc
Elizabeth	dbca
George	cabd
Martha	cdab

They noticed the following:
Andrea's and Bertha's add up to Dorothy's
Bertha's and Elizabeth's add up to George's
Caroline's and Dorothy's add up to Martha's

What is Andrea's extension?

77

QUID PRO QUO

Victor Bryant

In Readiland the unit of currency is the quid. Notes are available in two denominations and with these notes it is possible to make any three-figure number of quid. However, you need a mixture of the denominations to make exactly 100 quid. Furthermore, there is only one combination of denominations that will give a total of 230 quid.

What are the two denominations?

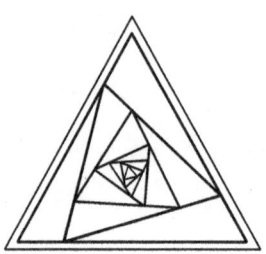

MR GREEN'S SCALENE MEAN MACHINE

Stephen Hogg

My maths teacher, Mr. Green, stated that the average of the squares of any two different odd numbers gives the hypotenuse of a right-angled triangle that can have whole-number unequal sides, and he told us how to work out those sides.

I used my two sisters' ages (different odd prime numbers) to work out such a triangle, then did the same with my two brothers' ages (also different odd prime numbers). Curiously, both triangles had the same three-figure palindromic hypotenuse. However, just one of the triangles was very nearly a 45° right-angled triangle (having a relative difference between the adjacent side lengths of less than 2%).

In ascending order, what were my siblings' ages?

DEBASED

Howard Williams

Sarah writes down a four-digit number then multiplies it by four and writes down the resultant five-digit number. She challenges her sister Jenny to identify anything that is special about these numbers. Jenny is up to the challenge as she identifies two things that are special. She sees that as well as both numbers being perfect squares she also recognizes that if the five-digit number was treated as being to base 7 it would, if converted to a base 10 number, be the same as the original four-digit number.

What is the four-digit number?

WILLIAM'S PRIME

Bill Scott

William was searching for a number he could call his own. By consistently replacing digits with letters, he found a number represented by his name: WILLIAM.

He noticed that he could break WILLIAM down into three smaller numbers represented by WILL, I and AM, where WILL and AM are prime numbers.

He then calculated the product of the three numbers WILL, I and AM.

If I told you how many digits there are in the product, you would be able to determine the number represented by WILLIAM.

What number is represented by WILLIAM?

BAZ'S BIZARRE ARRERS

Stephen Hogg

'Bizarrers' dartboards have double and treble rings and twenty sectors ordered as on this normal dartboard. However, a sector's central angle is [100 divided by its basic score]°. The 20 sector incorporates the residual angle to complete 360°.

Each player starts on 501 and reduces this, eventually to 0 to win. After six three-dart throws, Baz's seventh could win. His six totals were consecutive numbers. Each three-dart effort lodged a dart in each of three clockwise-adjacent sectors (hitting, in some order, a single zone, a double zone and a treble zone). The three-sector angle sum (in degrees) exceeded that total.

The sectors scored in are calculable with certainty, but not how many times hit with certainty, except for one sector.

Which sector?

FIELD FOR THOUGHT

Peter Good

Farmer Giles had a rectangular field bordered by four fences that was 55 hectares in size. He divided the field into three by planting two hedges, from the mid-point of two fences to two corners of the field. He then planted two more hedges, from the mid-point of two fences to two corners of the field. All four hedges were straight, each starting at a different fence and finishing at a different corner.

What was the area of the largest field when the four hedges had been planted?

SPORTY SET

Victor Bryant

There are 100 members of my sports club where we can play tennis, badminton, squash and table tennis (with table tennis being the least popular). Last week I reported to the secretary the numbers who participate in each of the four sports. The digits used overall in the four numbers were different and not zero.

The secretary wondered how many of the members were keen enough to play all four sports, but of course he was unable to work out that number from the four numbers I had given him. However, he used the four numbers to work out the minimum and the maximum possible numbers playing all four sports. His two answers were two-figure numbers, one being a multiple of the other.

How many played table tennis?

84

TIMELY COINCIDENCE

Danny Roth

George and Martha possess two digital 'clocks', each having six digits. One displays the time on a 24- hour basis in the format hh mm ss, typically 15 18 45, and the other displays the date in the format dd mm yy, typically 18 07 14.

On one occasion, George walked into the room to find that the two 'clocks' displayed identical readings. Martha commented that the long-term (400-year) average chance of that happening was 1 in just over a six-digit number. That six-digit number gives the birth date of one their daughters.

On what date was that daughter born?

TRIPLE JUMP

Bill Kinally

From a set of playing cards, Tessa took 24 cards consisting of three each of the aces, twos, threes, and so on up to the eights. She placed the cards face up in single row and decided to arrange them such that the three twos were each separated by two cards, the threes were separated by three cards and so forth up to and including the eights, duly separated by eight cards. The three aces were numbered with a one and were each separated by nine cards. Counting from the left, the seventh card in the row was a seven.

In left to right order, what were the numbers on the first six cards?

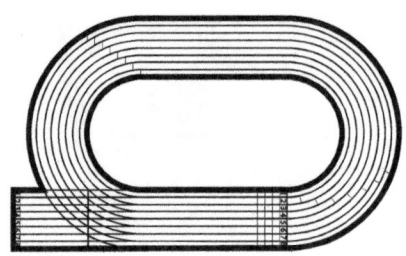

PLEASE MIND THE GAP

Howard Williams

Ann, Beth and Chad start running clockwise around a 400m running track. They run at a constant speed, starting at the same time and from the same point; ignore any extra distance run during overtaking.

Ann is the slowest, running at a whole number speed below 10 m/s, with Beth running exactly 42% faster than Ann, and Chad running the fastest at an exact percentage faster than Ann (but less than twice her speed).

After 4625 seconds, one runner is 85m clockwise around the track from another runner, who is in turn 85m clockwise around the track from the third runner.

They decide to continue running until gaps of 90m separate them, irrespective of which order they are then in.

For how long in total do they run (in seconds)?

PARTY TIME

Graham Smithers

A four-digit number with different positive digits and with the number represented by its last two digits a multiple of the number represented by its first two digits, is called a PAR.

A pair of PARs is a PARTY if no digit is repeated and each PAR is a multiple of the missing positive digit.

I wrote down a PAR and challenged Sam to use it to make a PARTY. He was successful.

I then challenged Beth to use my PAR and the digits in Sam's PAR to make a different PARTY. She too was successful.

What was my PAR?

88

LONG SHOT

Andrew Skidmore

Callum and Liam play a simple dice game together using standard dice (numbered 1 to 6). A first round merely determines how many dice (up to a maximum of three) each player can use in the second round. The winner is the player with the highest total on their dice in the second round.

In a recent game Callum was able to throw more dice than Liam in the second round but his total still gave Liam a chance to win. If Liam had been able to throw a different number of dice (no more than three), his chance of winning would be a whole number of times greater.

What was Callum's score in the final round?

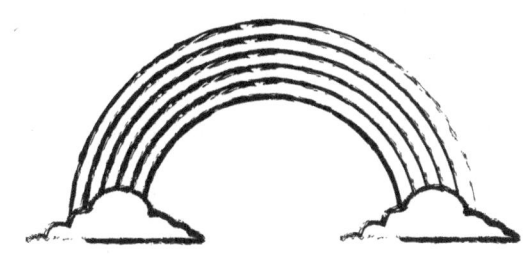

RAINBOW NUMERATION

Stephen Hogg

Dai had seven standard dice, one in each colour of the rainbow (ROYGBIV). Throwing them simultaneously, flukily, each possible score (1 to 6) showed uppermost. Lining up the dice three ways, Dai made three different seven-digit numbers: the smallest possible, the largest possible, and the 'rainbow' (ROYGBIV) value. He noticed that, comparing any two numbers, only the central digit was the same and each number had just one prime factor under 10 (different for each number).

Hiding the dice from his sister Di's view, he told her what he'd done and noticed, but wanted her to guess the 'rainbow' number digits in ROYGBIV order. Luckily guessing the red and orange dice scores correctly, she then calculated the others unambiguously.

What score was on the indigo die?

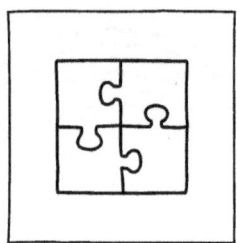

SQUARE JIGSAWS

Victor Bryant

I chose a whole number and asked my grandson to cut out all possible rectangles with sides a whole number of centimetres whose area, in square centimetres, did not exceed my number. (So, for example, had my number been 6 he would have cut out rectangles of sizes 1x1, 1x2, 1x3, 1x4, 1x5, 1x6, 2x2 and 2x3.) The total area of all the pieces was a three-figure number of square centimetres.

He then used all the pieces to make, in jigsaw fashion, a set of squares. There were more than two squares and at least two pieces in each square.

What number did I originally choose?

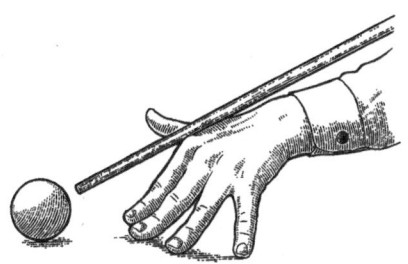

POT SUCCESS

John Owen

In snooker, pot success (PS) is the percentage of how many pot attempts have been successful in that match (e.g. 19 pots from 40 attempts gives a PS of 47.5). In a recent match, my PS was a positive whole number at one point. I then potted several balls in a row to finish a frame, after which my improved PS was still a whole number. At the beginning of the next frame, I potted the same number of balls in a row, and my PS was still a whole number. I missed the next pot, my last shot in the match, and, remarkably, my PS was still a whole number.

If I told you how many balls I potted during the match, you would be able to work out those various whole-number PS values.

How many balls did I pot?

92

END OF THE BEGINNING

Howard Williams

Jenny is using her calculator, which accepts the input of numbers of up to ten digits in length, to prepare her lesson plan on large numbers. She can't understand why the results being shown are smaller than she expected until she realizes that she has entered a number incorrectly.

She has entered the number with its first digit being incorrectly entered as its last digit. The number has been entered with its second digit first, its third digit second etc. and what should have been the first digit entered last. The number she actually entered into her calculator was 25/43rds of what it should have been.

What is the correct number?

DARTS DISPLAY

Andrew Skidmore

I noticed a dartboard in a sports shop window recently. Three sets of darts were positioned on the board. Each set was grouped as if the darts had been thrown into adjacent numbers (e.g. 5, 20, 1) with one dart from each set in a treble. There were no darts in any of the doubles or bulls.

The darts were in nine different numbers but the score for the three sets was the same. If I told you whether the score was odd or even you should be able to work out the score. The clockwise order of numbers on a dartboard is:

20 1 18 4 13 6 10 15 2 17 3 19 7 16 8 11 14 9 12 5

What was the score made by each set of three darts?

94

GOLDILOCKS AND
THE BEAR COMMUNE

Mike Fletcher

In the bears' villa there are three floors, each with 14 rooms. The one switch in each room bizarrely toggles (on ↔ off) not only the single light in the room but also precisely two other lights on the same floor; moreover, whenever A toggles B, then B toggles A.

As Goldilocks moved from room to room testing various combinations of switches, she discovered that on each floor there were at least two separate circuits, no two circuits on a floor had the same number of lights, and no two floors have the same combination of circuits. Furthermore, she found a combination of 30 switches that turned all 42 lights from 'off' to 'on', and on one floor she was able turn each light on by itself.

(a) How many circuits are there?
(b) How many lights are in the longest circuit?

95

RESERVOIR DEVELOPMENT

Graham Smithers

A straight track from an observation post, O, touches a circular reservoir at a boat yard, Y, and a straight road from O meets the reservoir at the nearest point, A, with OA then extended by a bridge across the reservoir's diameter to a disembarking point, B. Distances OY, OA and AB are whole numbers of metres, with the latter two distances being square numbers.

Following development, a larger circular reservoir is constructed on the other side of the track, again touching OY at Y, with the corresponding new road and bridge having all the same properties as before. For both reservoirs, the roads are shorter than 500m, and shorter than their associated bridges. The larger bridge is 3969m long.

What is the length of the smaller bridge?

FRIEND OF THE DEVIL

Stephen Hogg

My friend, 'Skeleton' Rose, rambled on with me and my uncle ('The Devil' and 'Candyman') about Mr. Charlie, who gave, between us, three identical boxes of rainbow drops.

Each identical box's card template had a white, regular convex polygonal base section with under ten sides, from each of which a similar black triangular star point extended. All these dark star points folded up to an apex, making a closed box.

The number of sweets per box equalled the single-figure sum of its own digits times the sum of the template's star points and the box's faces and edges. If I told you how many of the 'star point', 'face' and 'edge' numbers were exactly divisible by the digit sum, you would know this number of sweets.

How many sweets were there in total?

THAT OLD CHESTNUT

Colin Vout

Clearing out an old drawer I found a wrinkled conker. It was my magnificent old 6709-er, a title earned by being the only survivor of a competition that I had had with friends. The competition had started with five conkers, veterans of many campaigns; each had begun at a different prime value between 1300 and 1400.

We used the rule that if an m-er beat an n-er in an encounter (by destroying it, of course!) the m-er would become an m+n+1-er; in effect, at any time the value of a conker was the number of destroyed conkers in all confrontations in its 'ancestry'.

I recall that, throughout the competition, the value of every surviving conker was a prime number.

What were the values of the five conkers at the start?

98

PRIME ADVENT CALENDAR

John Owen

Last year I was given a mathematical Advent calendar with 24 doors arranged in four rows and six columns, and I opened one door each day, starting on the 1st December. Behind each door is an illustrated prime number, and the numbers increase each day. The numbers have been arranged so that once all the doors have been opened, the sum of the numbers in each row is the same, and likewise for the six columns. Given the above, the sum of all the prime numbers is as small as it can be.

On the 24th, I opened the last door to find the number 107.

In order, what numbers did I find on the 20th, 21st, 22nd and 23rd?

99

PROGRESSIVE RAFFLE

Danny Roth

George and Martha were participating in the local village raffle. 1000 tickets were sold, numbered normally from 1 to 1000, and they bought five each. George noticed that the lowest-numbered of his tickets was a single digit, then each subsequent number was the same multiple of the previous number, e.g. 7 21 63 189 567. Martha's lowest number was also a single digit, but her numbers proceeded with a constant difference, e.g. 6 23 40 57 74. Each added together all their numbers and found the same sum. Furthermore, the total of all the digits in their ten numbers was a perfect square.

What was the highest numbered of the ten tickets?

THREE PATTERNED PAVING

Howard Williams

James is laying foot-square stones in a rectangular block whose short side is less than 25 feet. He divides this area into three rectangles by drawing two lines, parallel to the shorter sides, and into each of these three areas he lays a similar pattern.

The pattern consists of a band or bands of red stones laid concentrically around the outside of the rectangles with the centre filled with white stones. The number of red stone bands is different in each of the rectangles but in each of them the number of white stones used equals the number of red stones used.

The total number of stones required for each colour is a triangular number (i.e. one of the form 1+2+3+...).

What is the total area in square feet of the block?

SOLUTIONS

1 LONG DIVISION

Answer: 37 and 58571

The 2 digit number has to be 37 and 3 has to be the last digit of the quotient.

The 2nd digit of my larger number has to be 8.

The 1st digit of my larger number cannot be 4 or 6, since each of them introduces an extra 1, and so this 1st digit has to be 5.

The 2nd digit in the quotient will be either a 4 or a 5. However, it cannot be a 4 since that would not yield the 8 in my larger number.

And so the quotient now reads 15*3 and so is either 1573, 1583 or 1593.

With 1573, long division of 58201 by 37 shows that 0, 2, 3, 5, 7, 8 and 9 have been erased, i.e. 7 erased.

With 1583, long division of 58571 by 37 shows that 0, 2, 3, 5, 6, 7, 8 and 9 have been erased, i.e. 8 erased.

With 1593, long division of 58941 by 37 shows that 2, 3, 4, 5, 7, 8 and 9 have been erased, i.e. 7 erased.

So 8 different digits were erased when dividing 58571 by 37.

2 BETTING TO WIN

Answer: 11-2 and £22

If the odds are X-Y, we can convert that to a "probability" of Y/(X+Y). Let the probability for a win for A be a, the probability for a win for C be b and the probability for a draw be c. The amount returned on a winning bet is the stake multiplied by the probability.

In order to guarantee to win a certain amount, then we bet na on A, nb on B and nc on a draw. In each case, we win n. The total outlay is na+nb+nc, so to make a profit of £1, we have n - n(a+b+c) = 1. Put another way, the sum of probabilities is (n-1)/n.

Now the odds for a win for A are 8-5, which gives a probability of 5/13

The odds for a win for B are 6-5, which gives a probability of 5/11

Then 5/13 + 5/11 + c = (n-1)/n, so c + 120/143 = (n-1)/n

Clearly, n must be a multiple of 143 for c to be a rational number Y/(X+Y) where X and Y are less than 20. If n=143, then c=2/13, so X=11 and Y=2. The odds are 11-2 and the amount bet on a draw is 143*2/13 or £22.

[To check for other values of n, let n=143m. Then we can rewrite our equation as c=(23m-1)/143m. We want 23m-1 to be divisible by 11 or 13. In the first case, m=1, 12, 23, ... and in the second case, m=4, 17, 30,... m=4 gives c=7/44 which isn't allowed. Larger values of m make c ever closer to 23/143 and it can be checked that no denominator for c in the range 1 to 40 will give a whole number for the numerator.]

3 BIG DEAL

Answer: 28 cards

With 52 cards the number of combinations is $52!/(13!)^4$

[There are 52! Ways of laying the pack in a row but the first 13 can be shuffled around, as can the second, third and fourth thirteens.]

With 4N cards the number of combinations is $(4N)!/(N!)^4$ [*] (always a whole number)

If this is a number of seconds then the number of days is * divisible by 60.60.24 or $2^7 3^3 5^2$ and so, in particular, * must be divisible by 2^7.

Let $A = (4N)!$ and $B = 2^7 3^3 5^2 (N!)^4$: we need A/B to be an integer. We shall start by counting the factors of 2 in A and B and tabulate the results below..

[So for N=1 we have A=24 and B=$2^7 3^3 5^2$. Thereafter, when increasing N by 1, A is multiplied by (4N+4)(4N+3)(4N+2)(4N+1) and B is multiplied by $(N+1)^4$]

N	2s in A	2s in B				
1	3	7	↓ A x 8.7.6.5		↓ B x 2^4	
2	7	11	↓ A x 12.11.10.9		↓ B x 3^4	
3	10	11	↓ A x 16.15.14.13		↓ B x 4^4	
4	15	19	↓ A x 20.19.18.17		↓ B x 5^4	
5	18	19	↓ A x 24.23.22.21		↓ B x 6^4	
6	22	23	↓ A x 28.27.26.25		↓ B x 7^4	
7	25	23	↓ A x 32.31.30.29		↓ B x 8^4	
8	31	35	↓ A x 36.35.34.33		↓ B x 9^4	
9	34	35	↓ A x 40.39.38.37		↓ B x 10^4	
10	38	39	↓ A x 44.43.42.41		↓ B x 11^4	
11	41	39	↓ A x 48.47.46.45		↓ B x12^4	
12	46	47				

In order to get a whole number of days, the 2s in A must be at least the number of 2s in B. This rules out the shaded pairs and just leaves N=7 or N=11. But in the case N=11 it is easy to check that B has a factor of 5^{10} whereas A only has a factor of 5^9, so we will not get a whole number of days.

It is easy to check that N=7 does work:

Factors of 3: A has 11, B has 9

Factors of 5: A has 6 and B has 6.

So N=7 and the pack has 28 cards.

4 WHAT DO POINTS MAKE?

Answer: BvA 1-2 BvC 2-2 BvD 1-0

The points add up to 15 and all are different. Six games are played in total, so three of them must be draws (with 6 points scored in total), while 9 points are scored in the other three games. If D loses all its games (scoring no points and no goals), then the other three games played must all be draws. In that case, A, B and C will all end up with the same points (5), which isn't allowed. Therefore, D must score at least 1 goal.

9, 3, 2, 1 requires WWW, DDD or WLL, LDD, DLL.
WWW is incompatible with DDD but the WLL option involves three Ds but there has to be an even number.
8, x, y, z is impossible because a team cannot score 8 points in 3 games.
7, 5, 2, 1 and 7, 4, 3, 1 are viable.
6, 5, 3, 1 requires WWL, WDD, WLL or DDD, DLL.
DDD is incompatible with WWL; but WLL involves three Ds.
6, 4, 3, 2 involves WWL, WDL, WLL or DDD, DDL.
DDD is incompatible with WWL; but WLL involves three Ds.

So the points distributions and records could be
a) 7, 5, 2, 1 WWD, WDD, DDL, DLL
b) 7, 4, 3, 1 WWD, WDL, DDD, DLL
c) 7, 4, 3, 1 WWD, WDL, WLL, DLL We can rule this out because it involves three Ds.

One of the games was 5-0. Those 5 goals can't have been scored by B in case a) because B must have scored at least one goal in the two draws, which cannot both be 0-0. So one of A's results was 5-0.

a) 7, 5, 2, 1 WWD, WDD, DDL, DLL
    a.i)   A: 5-0, 2-0/2-1, 0-0 The 0-0 draw must be with undefeated B. But now C's draws are 1-1 and 2-2 which gives C at least 3 goals.
    a.ii) A: 5-0, 1-0, 1-1; immediately BvA 1-1 CvB 2-2, CvD 0-0. Now BvD has to be 2-0 (it can't be 2-1) to give B 5 goals, but now D's three games are 0-0, 0-2 and 0-1 or 0-5 scoring no goals.

b) 7, 4, 3, 1 WWD, WDL, DDD, DLL
    b.i)   A: 5-0, 2-0 or 2-1, 0-0;
            immediately CvA 0-0, CvD 1-1, CvB 2-2; D loses the other two games 0-x, 0-y
    b.i.1) If DvA 0-5 and AvB 2-0 then BvD has to be 2-0 (not 2-1) to get B to 4 goals but this repeats 2-0.
    b.i.2) If DvA 0-5 and AvB 2-1 then BvD 1-0 to get B to 4 goals. This gives the solution.
            AvB 2-1 AvC 0-0 AvD 5-0
            BvA 1-2 BvC 2-2 BvD 1-0
            CvA 0-0 CvB 2-2 CvD 1-1
            DvA 0-5 DvB 0-1 DvC 1-1
    b.i.3) If DvA 0-2 then BvA 0-5, CvB 2-2 and now BvD is 2-0 (not 2-1) repeating 2-0.
    b.ii) A: 5-0, 1-0, 1-1
            Immediately D 0-0 1-x 0-5 then B 2-2, x-1 0-1 so x = 2 but now BvD has 3 goals.

5 KEEP YOUR BALANCE

Answer: 4, 8, 5, 9, 12, 11, 1, 6, 2, 7, 10, 3

The first step is to establish the range of prime numbers. The lowest total possible of a threesome is $1 + 2 + 3 = 6$ so 7 is theoretically the lowest possible prime and the highest is $10 + 11 + 12 = 33$ so the highest theoretically possible prime is 31. Thus the following are the eight candidates:

7 11 13 17 19 23 29 31. Now each total has to be satisfied twice. 7 is thus ruled out with $1 + 2 + 4$ but no alternative. 31 can similarly be eliminated as it is $8 + 11 + 12$ or $9 + 10 + 12$ so that we cannot have a weighing without a repetition. Similarly for 29, we can have $8 + 10 + 11$ or $8 + 9 + 12$ or $7 + 10 + 12$ or $6 + 11 + 12$. Again no weighing is possible without repetition. So we are left with 11 13 17 19 and 23.

Now weighing   1) will total 22
              2)   26
              3)   34
              4)   38
              5)   46

That gives us a total of 166. We notice that each letter appears at least twice (24 so far) and that implies a total so far of $12 \times 13 = 156$, leaving 10 unaccounted for with six remaining. Any number can only appear a maximum of five times without repetition so that we can only have three extras for any one number. We thus need $1 + 1 + 1 + 2 + 2 + 3$; there appears to be no other way.

Thus we have five 1's, four 2's, three 3's and two of each of the others. We observe that G appears five times so that $= 1$ and that I appears four times so that must be 2. Similarly L appears three times so that must be 3. Thus we can rewrite the five equations with the information we have so far:

         1) $A + C + 2$ v $1 + J + 3$ total 11 so $J = 7$
         2) $A + H + 3$ v $1 + 2 + K$ total 13 so $K = 10$
         3) $B + 2 + J$ v $C + F + 1$
         4) $B + D + 2$ v $E + 1 + H$
         5) $D + F + 3$ v $E + 1 + K$

Entering the information from the above calculations:

         1) $A + C + 2$ v $1 + 7 + 3$ total 11 so $J = 7$
         2) $A + H + 3$ v $1 + 2 + 10$ total 13 so $K = 10$
         3) $B + 2 + 7$ v $C + F + 1$ total 17 so $B = 8$
         4) $B + D + 2$ v $E + 1 + H$
         5) $D + F + 3$ v $E + 1 + 10$ total 23 so $E = 12$

continuing this argument:

         1) $A + C + 2$ v $1 + 7 + 3$ total 11 so $J = 7$
         2) $A + H + 3$ v $1 + 2 + 10$ total 13 so $K = 10$
         3) $8 + 2 + 7$ v $C + F + 1$ total 17 so $B = 8$
         4) $8 + D + 2$ v $12 + 1 + H$ total 19 so $H = 6$
         5) $D + F + 3$ v $12 + 1 + 10$ total 23 so $E = 12$

and again:

1) $A + C + 2$ v $1 + 7 + 3$ total 11 so J = 7
2) $A + 6 + 3$ v $1 + 2 + 10$ total 13 so K = 10 and A = 4
3) $8 + 2 + 7$ v $C + F + 1$ total 17 so B = 8 and with F = 11
                                                   as below C = 5
4) $8 + D + 2$ v $12 + 1 + 6$ total 19 so H = 6 and D = 9
5) $D + F + 3$ v $12 + 1 + 10$ total 23 so E =12 and F = 11

So we have A = 4, B = 8, C = 5, D = 9, E = 12, F = 11, G = 1, H = 6, I = 2, J = 7, K = 10, L = 3

6 HAPPY FAMILIES

Answer: George, Kate and Larry

Using initials

	1	2	3	4	5
Father	J	A	G	D	M
Mother	H	B	E	K	N
Child	F	I	C	L	O

Since B's number is one less than C's, and G's one less than L's, neither C nor L is at no 1. Similarly, neither B nor G is at no 4. The mother at no 3 is not B or H. D is not at no 1 nor is I. So, the mother at no 3 is either Ellen or Kate. The best way to proceed is now by trial and error, which eventually gives the answer above.

Those in the correct positions are George, Kate and Larry.

Answer: 17

Number of battles=lowest number not a factor of [6^(6^6)]-[(6^6)^6]

Now 6^6=46656, so 6^(6^6) = 6^(36+46620) = [6^36][6^46620]

Also, (6^6)^6 = 6^36, so [6^(6^6)]-[(6^6)^6] = [6^36][(6^46620)-1]

Factors of 6^36=[2^36][3^36] are 2,3,4,6,8,9,12,16,18,24 etc

Factors of [(6^46620)-1] include 5 (6^N ends 6 for N integer>0) and with above 10, 15, etc

Check [(6^46620)-1] for prime factors 7, 11, 13, 17, etc to fill gaps (and valid multiples)

Look at the pattern of remainders of 6^N on division by possible prime factors P for values of N (the pattern will repeat once you have again got to a remainder of 1):

N	0	1	2	3	4	5	6	7	8	9	10	11	12	13	14	15	16
P																	
7 {6^N mod 7}	1	6	1														
11 {6^N mod 11}	1	6	3	7	9	10	5	8	4	2	1						
13 {6^N mod 13}	1	6	10	8	9	2	12	7	3	5	4	11	1				
17 {6^N mod 17}	1	6	2	12	4	7	8	14	16	11	15	5	13	10	9	3	1

If N=46620=2x2x3x3x5x7x37, then

P=7, N is divisible by 2 and {6^N mod 7} =1, so 7 is a factor of [(6^N)-1] (and 14, etc.)
P=11, N is divisible by 10 and {6^N mod 11}=1, so 11 is a factor of [(6^N)-1]
P=13, N is divisible by 12 and {6^N mod 13}=1, so 13 is a factor of [(6^N)-1]
P=17, N is not divisible by 16, so {6^N mod 17} is not 1 and 17 is not factor of [(6^N)-1]

8 PENTAGONAL GARDENS

Answer: 16384

Let the sides be x so the area of lawn is x^2. If there are two right-angled corners they are either together or separated by another corner.

If they are together the shape of the garden is like this:

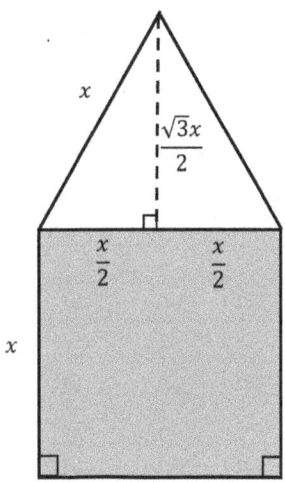

Area: $x^2 + \frac{\sqrt{3}}{4}x^2$ Paved area: $\frac{\sqrt{3}}{4}x^2$

If the right-angled corners are separated the garden looks like this (rotate the right-hand isosceles triangle until the fifth side is x)

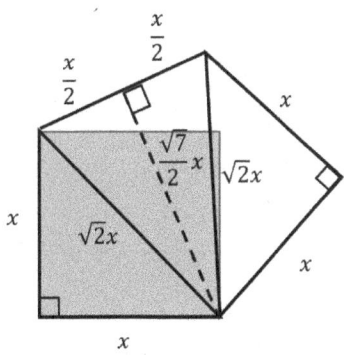

Area: $x^2 + \frac{\sqrt{7}}{4}x^2$ Paved area: $\frac{\sqrt{7}}{4}x^2$

The difference of the areas multiplied by the sum of the paved areas is

$$\left(\frac{\sqrt{7}}{4}x^2 - \frac{\sqrt{3}}{4}x^2\right) \times \left(\frac{\sqrt{7}}{4}x^2 + \frac{\sqrt{3}}{4}x^2\right) = \frac{7}{16}x^4 - \frac{3}{16}x^4 = \frac{x^4}{4}$$

The only value of x which gives a five-digit number with five different digits is 16, and the answer is 16384.

9 55 DIVISIONS

Answer: 143869275

55 = 5*11! To be divisible by 5 the last digit of the long number must be 5.

To be divisible by 11 the difference between the sum of the 1st, 3rd, ... etc and the 2nd, 4th ...etc digits must be zero or a multiple of 11. The total of digits 1 to 9 is 45 and therefore one of the sums must be 28 and the other 17 – ie 28+17=45 and 28–17 = 11.

There are two strings of digits, one of them having 5 digits and the second 4 digits. Totals are 17 and 28 and in either order. The digit '5' must be in the longer string as it has to be the last digit of the 9- and, of course, of the 6-digit and the 3-digit number. The possibilities for the strings of digits are:

	Short string total 17	Long string total 28
A	1, 2, 6, 8	3, 4, 7, 9 5
B	1, 3, 6, 7	2, 4, 8, 9 5
C	1, 3, 4, 9	2, 6, 7, 8 5
D	2, 3, 4, 8	1, 6, 7, 9 5
	Short string total 28	**Long string total 17**
E	4, 7, 8, 9	1, 2, 3, 6 5

The 9-digit number must be made up as follows: L S L S L S L S 5

L is a digit from the long string and S from the short string. The gaps in the string in the above show the delineation of the 3 digits which are dropped in two steps.

Let the three 3-digit numbers be P, Q and R; the total (T) is $P*10^6 + Q*10^3 + R$ which is a multiple of 11. Removing the first 3 digits is the equivalent of subtracting $P*10^6$ from T. $T - P*10^6$ is also a multiple of 11, and so P must also be a multiple of 11. Similarly Q and R must be multiples of 11. Therefore we are looking for three 3-digit multiples of 11 in the form above. The answer must be the smallest possible and therefore, as a start, we should only look at 9-digit numbers which begin with '1' – ie D & E above are possibles. These are possible P, Q and R for D & E:

			LSL	SLS	LS5
D	2, 3, 4, 8	1, 6, 7, 9 5	187, 627, 649	264, 374, 418	935
E	4, 7, 8, 9	1, 2, 3, 6 5	143, 176, 286, 396	418, 429, 869	275, 385

Therefore by trial and error there are three 9-digit numbers beginning with 1:

E: 143 869 275   NB The 8 and the 9 could be swapped, but not for the smallest number

E: 176 429 385   NB As above for 4 and 9

D: 187 264 935   NB As above for 2 and 4

Therefore the answer is 143869275.

Answers: (a) 2015 (b) 523

In 2019 the digit sum is 6 short of a multiple of 9. The same must be true of the sum of the eight different non-zero digits used and so if the sum is possible it will not use a 6.

Likewise if the 2018 sum is possible it will not use a 7.

If the 2017 sum is possible it will not use a 8.
If the 2016 sum is possible it will not use a 9.
If the 2015 sum is possible it will not use a 1.
If the 2014 sum is possible it will not use a 2.
If the 2013 sum is possible it will not use a 3.

But we can soon see that all the sets of numbers must include a 4-figure number of the form 1--- and so it is impossible to make the total **2015** in the required way.

[All others are possible going back for over 10 years: e.g.

2007=8+657+1342   *2008=9+657+1342*   *2009=9+658+1342*

2010=9+578+1423   *2011=9+678+1324*   *2012=9+678+1325*

2013=9+578+1426   *2014=9+568+1437*   *2016=8+576+1432*

2017=9+576+1432   *2018 & 2019 – see below]*

In 2018 the sum uses the digits 12345689 and it includes a 3-figure prime (hence ending in 3 or 9). So the numbers are of the form

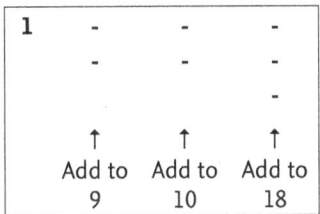

1	-	-	-
	-	-	-
			-
↑	↑	↑	
Add to 9	Add to 10	Add to 18	

→ With digits in columns in no particular order

1	4	2	3
5	8	6	
		9	

or

1	3	2	4
6	8	5	
		9	

The equivalent answers for 2019 (again where digits in columns can be shuffled) are

1	4	3	2
5	7	8	
		9	

or

1	4	2	3
5	8	7	
		9	

The 3-figure numbers ending 3 or 9 that can occur in both years are

423   429   483   489   523   529   583   589

All but one of these has a factor as shown:

423(3)   429(3)   483(3)   489(3)   **523**   529(23)   583(11)   589(19)
                                    prime

[Possible solutions are 2018=9+523+1486 and 2019=9+523+1487]

Answer: 23 and 33603

The basic grid can be written as:

			p	q	r	s
a	#	(	#	#	b	c	#
			d	#		
			e	f	b	
			h	i		
			j	k	c	
			m	#	n	
				x	#	
				x	#	

We cannot have # = 0 or 1

If # = 2, then p = 1, d = 1, a = 1, but cannot find r to give the line m # n.

If # = 4, then p = 1, d = 3, a = 3, but cannot find r to give the line m 4 n.

If # = 5, then e = 1, d = 4 and either p = 1, a = 4 or p = 3, a = 1. But with either 45 or 15 as the 2-digit number, we cannot find r to give the line m 5 n.

If # = 6, then p = 1, a = 5, q = 1, which means i = 6 = #, which is impossible.

If # = 7, then p = 1, d = 6, a = 6, q = 1, which means i = 7 = #, which is impossible.

If # = 8, then p = 1, d = 7, a = 7, q = 1, which means i = 8 = #, which is impossible.

If # = 9, then p = 1, d = 8, a = 8, q = 1, which means i = 9 = #, which is impossible.

If # = 3, then f = 0, d = 2, e = 1, p = 1, a = 2. Then r multiplied by 23 gives the line m 3 n if r = 6, m = 1 and n = 8. That gives s = 1, x = 2, c = 0, k = 4, j = 1, h = 9, q = 4, i = 2 and b = 6. And so the division sum is 33603 divided by 23.

Answer: 8778

We must have   - x - - x - - - = - - - -

The smallest possible products when zero is not used are given by 356x2x14 = 9968 and 357x2x14 = 9996, neither of which work. All others have more than four digits.

The zero must be the middle digit of the three-digit number to avoid a zero in the product.

There are four possible cases –

a) 10X x 2Y x 3   c) 20X x 1Y x 3

b) 10X x 3Y x 2   d) 30X x 1Y x 2   (405 x 13 x 2 = 10530)

Choices for X and Y are limited by the 'unit digit' generated in the product being unable to have values of 0, 1, 2 or 3. Consider the 'unit digit' formed by 3xXxY in a) and 2xXxY in b) –

a) 3xXxY	→	XxY	→	X,Y
4		8		6,8
6		2		4,8
6		2		8,9
8		6		4,9

b) 2xXxY	→	XxY	→	X,Y
4		2		6,7
4		2		8,9
6		3		7,9
6		8		4,7
8		4		4,6
8		4		6,9

In c) and d) we are limited to Y < 7 on product size grounds ie Y = 4 or 6 (as 5 → 0). Possible cases are –

a) 106 x 28 x 3 = 8904
    108 x 26 x 3 = 8424
    104 x 28 x 3 = 8736
    108 x 24 x 3 = 7776
    108 x 29 x 3 = 9396
    109 x 28 x 3 = 9156
    104 x 29 x 3 = 9048
    109 x 24 x 3 = 7848

b) 106 x 37 x 2 = 7844
    107 x 36 x 2 = 7704
    108 x 39 x 2 = 8424
    109 x 38 x 2 = 8284
    107 x 39 x 2 = 8346
    109 x 37 x 2 = 8066
    104 x 37 x 2 = 7696
    107 x 34 x 2 = 7276
    104 x 36 x 2 = 7488
    106 x 34 x 2 = 7208
    106 x 39 x 2 = 8268
    109 x 36 x 2 = 7848

c) 20X x 1Y x 3

8	4	8736
9	4	8778*
8	6	9984

d) 30X x 1Y x 2

6	4	8568
7	4	8596
4	6	9728
7	6	9824
9	6	9888

The product is 8778   (209 x 14 x 3 = 8778)

Answer: 3/140

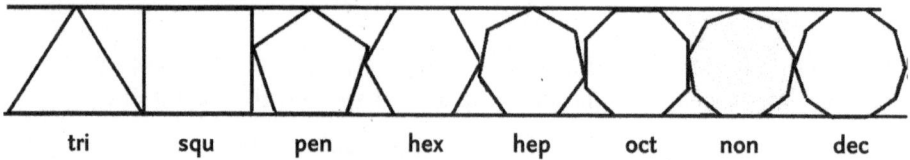

| tri | squ | pen | hex | hep | oct | non | dec |

All prisms are same height on a flat surface, as shown. For both triads, one overhangs both others and one is overhung by both others – so A overhangs B, each overhanging C. These constraints and regular polygon 'visual' symmetries mean:-

*the **tri** prism can't overhang any other.
*the **squ** prism excluded – can't overhang or be overhung by any other.
*the **oct** prism excluded – isn't overhung by any other and only overhangs the **tri** prism.
*the **hex** and **dec** prisms, as a pair, excluded in a triad, widest points are both at half-height.
*only the **tri**, **hex**, **hep** and **dec** prisms can be overhung by two others.

So for permissible triads (< means **'is overhung by'**) we get:

tri<[hex<pen]; tri<[hep<pen]; tri<[non<pen]; tri<[dec<pen]; tri<[hep<hex]; tri<[hex<non];
tri<[hep<non]; tri<[dec<non]; tri<[hep<dec];
hex<[non<pen];
hep<[hex<pen]; hep<[non<pen]; hep<[dec<pen]; hep<[hex<non]; hep<[dec<non];
dec<[non<pen]

Both triads meet the 'overhang' criteria, so need pairs from above not repeating a prism viz.

tri<[hex<pen] and hep<[dec<non]	tri<[dec<pen] and hep<[hex<non]
tri<[hep<hex] and dec<[non<pen]	tri<[hex<non] and hep<[dec<pen]
tri<[dec<non] and hep<[hex<pen]	tri<[hep<dec] and hex<[non<pen]

1st triad is one of these 12 from all possible triads from 8 prisms: 8!/(3!5!)=56 triads.
So **chance of permissible 1st triad is 12/56=3/14**
2nd triad must be only one without the **'squ'/'oct'** prisms (5 prisms: 5!/(3!2!)=10 triads)
So **chance of permissible 2nd triad is 1/10**
So **overall chance is (3/14)x(1/10)=3/140 (in simplest form)**

*For purists and completeness, by trig. :-
Ratio of height of widest part of regular odd-n-gon from a side to full height
(Inradius+Circumradius) = $h/H=[\cos(\pi/n)+\{(-1)^{\wedge}(\{n-1\}/2)\}\cos(\pi\{n-1\}/2n)]/[1+\cos(\pi/n)]$

So **tri**=$[\cos(\pi/3)-\cos(\pi/3)]/[1+\cos(\pi/3)]=0$; **pen**=$[\cos(\pi/5)+\cos(2\pi/5)]/[1+\cos(\pi/5)]\approx0.618$
hep=$[\cos(\pi/7)-\cos(3\pi/7)]/[1+\cos(\pi/7)]\approx0.357$; **non**=$[\cos(\pi/9)+\cos(4\pi/9)]/[1+\cos(\pi/9)]\approx0.574$

For others: **squ**=[0 to 1] range; **hex** and **dec**=0.5; **oct**=$[(1-1/\sqrt{2})\approx0.293$ to $1/\sqrt{2}\approx0.707]$ range

14 SILLY SLIP

Answer: 106496

	(1)	(2)	(3)	(4)	(5)	(6)
		S	I	L	L	Y
			S	L	I	P
	P	L	E	A	S	E

If all the Ls are correct then (from the L in column (2)) all the Ls are 0. Then the "A" in column (4) will be 0 or 1. If it's 0 (=L) then there is the slip. If it's 1 then it clashes with the fact that the "P" in column (1) is also 1. So if all the Ls are correct then the slip is at A or P. Therefore the slip is in one of the shaded positions below (numbered 1 to 6), the unshaded letters being correct:

		S	I	L³	L⁶	Y
			S	L⁴	I	P
	P¹	L²	E	A⁵	S	E

We consider the six possibilities in turn:

1:

	9	I	0	0	Y
		9	0	I	P≠1
1	0	E	A	9	E

Here A=0=L, a second slip

2:

	9	I	L≠0	L	Y
		9	L	I	1
1	0	E	A	9	E
1	0	0	0		

So (from the right) Y+1=E, L+I=9, A=2L, and I=E+1. We consider the possible values:

P	S	Y	E	I	L	A
1	9	0	1x			
1	9	2	3	4	5	10x
1	9	3	4	5	4x	
1	9	4	5	6	3	6x
1	**9**	**5**	**6**	**7**	**2**	**4**
1	9	6	7	8	1x	
1	9	7	8	9x		
1	9	8	9x			

The only case that works is shown in **boldface** and it gives

$$
\begin{array}{r}
9\,7\,2\,2\,5 \\
9\,2\,7\,1 \\
\hline
1\,0\,6\,4\,9\,6
\end{array}
$$

3/4:

	9	I	?	0	Y
		9	0	I	1
1	0	E	A	9	E

I=9=S

5:

	9	I	0	0	Y
		9	0	I	1
1	0	E	?	9	E

I=9=S

6:

	9	I	0	?	Y
		9	0	I	1
1	0	E	A	9	E

A=0=L or A=1=P

So there is only one case with just one slip, giving the answer as 106496

15 MARBLE TOWER

Answer: 20

The number of marbles needed for a layer is $1+2+...+x = x(x+1)/2$, where x is the number of marbles along the edge of the triangle. Now consider the position before the last bag is used.

If there were just 3 completed layers the number of marbles used would be $6 + 10 + 15 = 31$.

As 31 is not a multiple of 3 we know that $3 + 1 = 4$ bags cannot be correct.

Numerical solution

Number of layers(n)	Number of marbles used so far	Total	Number in last bag
2	3+6	9	1*
3	6+10+15	31	3
4	10+15+21+28	74	6
5	15+21+28+36+45	145	10
6	21+28+36+45+55+66	251	15
7	28+36+45+55+66+78+91	399	21

*can be neglected since there is more than one marble in each bag

$399 = 21 \times 19$, then 19 bags have been used so far, so there are 20 bags, with 21 marbles in each bag.

Algebraic solution

If Sn is the sum for n layers, Sn+1 for n+1 layers,

$Sn = \dfrac{n(n+1)+(n+1)(n+2)+.......2n(2n-1)}{2} = A+Bn+Cn^2+Dn^3+....$

$Sn+1 = \dfrac{(n+1)(n+2)+.... 2n(2n-1)+2n(2n+1)+(2n+1)(2n+2)}{2} = A+B(n+1)+C(n+1)^2+D(n+1)^3+...$

$Sn+1 - Sn = \dfrac{2n(2n+1)+(2n+1)(2n+2) - n(n+1)}{2} = B +C(2n+1)+D(3n^2+3n+1)$

→ $D = 7/6$, $C = 0$, $B = -1/6$ and using n=1 in Sn shows $A = 0$
$Sn = n/6[7n^2 - 1]$

The last layer needs $n/2[n-1]$ marbles hence $n/6[7n^2 - 1] = n/2[n-1](N)$ where N is the number of bags used so far.

$$N = \dfrac{7n^2 - 1}{3(n-1)}$$

Searching for integral values of N is quicker than a quadratic solution –

n	N	
2	27/3	9 (neglected)
3	62/6	
4	111/9	
5	174/12	
6	251/15	
7	342/18	19

There were 20 bags of marbles.

Checking many more values of n gives no whole number values for N and when n gets large, N gets closer and closer to (but never reaches) a whole number divided by 3, so there are no other whole number values for N.

17 GO FORTH AND MULTIPLY

Answer: 32768

a) If two of the right angles are adjacent in the hexagon then the other three sides must be as below, where the shaded square is fixed while the rhombus is variable. For convexity the rhombus would have to be a square, with $\theta = 90°$ but then the hexagon becomes a rectangle.

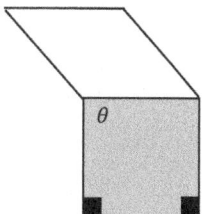

b) If two of the right angles are separated by another vertex the hexagon is as below left.

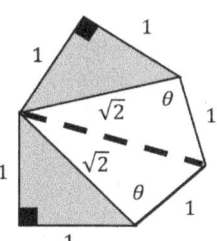
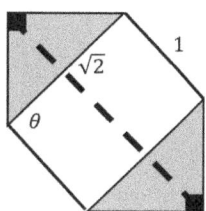

The area of the hexagon is the area of the fixed shaded triangles plus the area of the white kite, which is maximised when $\theta = 90°$ for a total area of $A = 1 + \sqrt{2}$.

c) Finally the two right angles are opposite each other in the hexagon above right. The area is maximised when the white parallelogram is a rectangle, also giving a total area of $A = 1 + \sqrt{2}$.

So the shapes of the two maximum-area gardens are as above, with paths on the axis of symmetry joining corners.

If the sides of the gardens are x then the lengths of the paths are $\sqrt{3}x$ and $(1 + \sqrt{2})x$. x is a natural number. Adam's calculation is:

$$A = \left((1 + \sqrt{2})x - \sqrt{3}x\right)\left((1 + \sqrt{2})x + \sqrt{3}x\right) = x^2(1 + \sqrt{2} - \sqrt{3})(1 + \sqrt{2} + \sqrt{3}) = 2\sqrt{2}x^2$$

Eve's calculation is $E = A^2 = \left(2\sqrt{2}x^2\right)^2 = 8x^4$. Let $x = 2^n k$ where k is odd.

$E = 8x^4 = 2^3(2^n k)^4 = 2^{4n+3}k^4$ is a perfect fifth power with no repeated digits.

So k is itself a perfect fifth power and $n = 3, 8, 13, ...$

The cases $k \neq 1$ do not concern us since the smallest value of E would be when $k = 3^5$, $n = 3$

$E = 2^{15} \times (3^5)^4 = 2^{15} \times (3^2)^{10} > 2^{15} \times (2^3)^{10} = 2^{45} > (2^{10})^4 > (1024)^4 > (1000)^4 = 10^{12}$.

This value of E would thus have at least 13 digits, and at least one of them must be repeated. For the cases $k = 1$, $n = 3, 8, 13$ ...

$n = 3$: $E = 2^{15} = 32768$, which is the answer.

$n = 8$: $E = 2^{35} = 2^5 \times (2^{10})^3 > 32 \times (1000)^3 = 3.2 \times 10^{10}$.

This value of E thus has at least 11 digits, at least one of which is repeated, and we need not consider larger n.

18   A NICE LITTLE EARNER

Answer: £249

[C(n) = net total handed out up to and including n, so -C(n) is the amount raised.]

For n < 86, C(n) > 0.

N: 86 87 88 89 90 91 92 93 94 95 96 97 98 99 100 101 102 103

-C(n). 0 1 1 0 9 17 24 30 35 39 42 44 45 45 46 46 45 43

N: 104 105 106 107 108 109 110 111 112 113 114 115

-C(n). 40 36   31 25 18 10 11 11 10 8   5   1

For n between 116 and 882 C(n) > 0.

{For easy counting note that 101 – 110 and subsequent intervals of 10 up to 181 – 190 add 35 each time; interval 191 – 200 adds 34; intervals from 201 – 210 up to 281 – 290 add 25 each time; interval 291 – 300 adds 24; and so on with additions of 15, 5, - 5, - 25 and – 35, until eventually you reach C(883) = - 1.}

You then get 2-digit negative sums up to C(910) = - 99. Then:

n: 911 912 913 914 915 916 917 918 919 920 921 922 923 924

-C(n): 107 114 120 125 129 132 134 135 135 144 152 159 165 170

   n: 925 926 927 928 929 930 931 932 933 934 935 936 937 938

-C(n): 174 177 179 180 180 189 197 204 210 215 219 222 224 225

   n: 939 940 941 942 943 944 945 946 947 948 949 950 951 952

-C(n): 225 234 242 249 255 260 264 267 269 270 270 279 287 294

   n: 953 954 955 956 957 958 959 960 961 962 963 964 965 966

-C(n): 300 305 309 312 314 315 315 324 332 339 345 350 354 357

   n: 967 968 969 970 971 972 973 974 975 976 977 978 979 980

-C(n): 359 360 360 369 377 384 390 395 399 402 404 405 405 414

   n: 981 982 983 984 985 986 987 988 989 990 991 992 993 994

-C(n): 422 429 435 440 444 447 449 450 450 459 467 474 480 485

  n: 995 996 997 998 999

-C(n): 489 492 494 495 495

And so C(942) = -£249, giving a profit of £249 with tickets numbered from 1 to 942.

Answer: 11 acres

Quadrilateral has unequal sides and so is not a parallelogram. Line through Leon and Anne resting is the 'Newton line' joining non-coincident mid-points of diagonals.

Signpost lies on the Newton line and fences from each corner to signpost make unequal triangles that conform to **'Anne's theorem'** in quadrilateral geometry.

**Sum of areas of two opposite triangles ('Three-acre meadow' and 'Elfland lea') =
Sum of areas of the other two opposite triangles ('Drax sward' and 'Plunkett's bawn') =
Half of total area of quadrilateral**

Shown with special case of a trapezium as follows:-

X is midpoint of a diagonal. In this case the Newton line is parallel to and midway between parallel sides AB and CD.

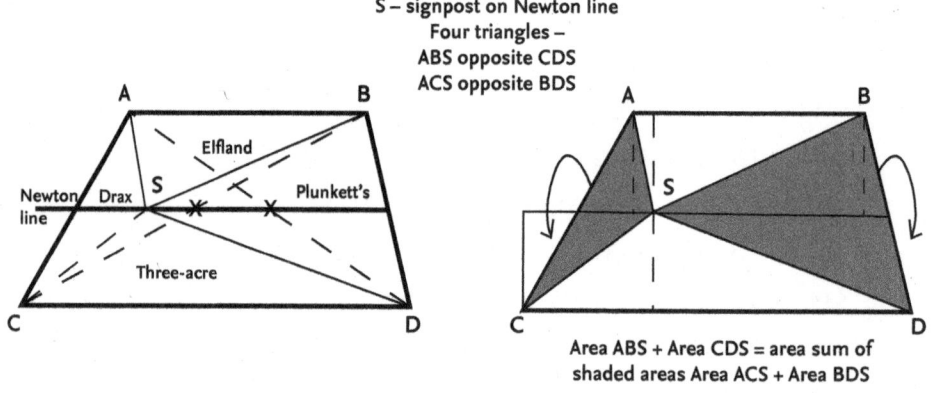

S – signpost on Newton line
Four triangles –
ABS opposite CDS
ACS opposite BDS

Area ABS + Area CDS = area sum of shaded areas Area ACS + Area BDS

So combined area of 'Three-acre meadow' (literally 3 acres) and 'Elfland lea'=3+E acres= combined area of 'Drax sward' and 'Plunkett's bawn'=Half area of 'Dunsany levels'

'Dunsany levels' area is whole number value=D=2(3+E)=2E+6

but 2<E<3, so 10<D<12 acres. Therefore, D=11

Answer: 1449

To start with, a perfect square of three different digits must end in 1 4 5 6 or 9. Furthermore, if 5 is applicable the previous digit must be a 2 and only 625 works as 225 is obviously out. If 1, 4 or 9 are applicable, the previous digit must be even and if 6 is applicable, the previous digit must be odd.

A prime number of three digits must end in 1, 3, 7 or 9. Thus 2 and 8 are absent on both counts and must both appear in the top left-hand corner, the adjacent squares or the middle. Thus at least one of them must be the most significant digit of a prime or perfect square. Thus the following are candidates to appear:

Primes: 239 241 251 257 263 269 271 283 293 821 823 827 829 839 853 857 863

Squares: 256 289 841

We know that at least one of those three perfect squares will be there. The best starting point is to establish whether 625 will be there as the second perfect square. That assumes that exactly one of the above three will be present. Possibilities are therefore as follows:

68 86   In both cases, we are left with 3 7 and 9. On the left, 687 divides by 3, 247 divides by 13 and 517 divides by 11.

24 42   On the right, 867 divides by 3; 427 divides by 7 and 157 is prime but 423 and 429 both divide by 3.

51 15

So 625 is eliminated and thus 5 also has to be barred from being the last digit of any of the numbers and will be on with 2 and 8 in or near the top left-hand corner. Candidates are thus reduced to:

25   52 28 82 85 58   2   2   5   5   8   8   2   2   5   5   8   8

8   8   5   5   2   2   58   85   28   82   25   52   58   85   28   82   25   52

It is now trial and error but there will be soon only one solution (with 643 being the diagonal prime):

283

547

619

The sum of the three "across" numbers is 1449

Answer: (a) HERN and RUSSELL (b) AVERY and DE HOYOS

First note that the only possible companion for NAVARRO is BRYNNER. Similarly, the only possible companion for WALLACH is VAUGHN. Given this, the only possible companion for AVEREY is DEXTER (since AVERY and WALLACH can't go together). Proceeding in this way, we eventually find that there are three possible solutions:

BRONSON	BISSELL & LUCERO
BRYNNER	NAVARRO & RUSKIN
BUCHHOLZ	BRAVO & SUAREZ
COBURN	ALONZO & VACIO
DEXTER	AVERY & DE HOYOS
MCQUEEN	HERN & RUSSELL
VAUGHN	ALANIZ & WALLACH

BRONSON	BISSELL & DE HOYOS
BRYNNER	NAVARRO & RUSKIN
BUCHHOLZ	BRAVO & SUAREZ
COBURN	ALONZO & VACIO
DEXTER	AVERY & LUCERO
MCQUEEN	HERN & RUSSELL
VAUGHN	ALANIZ & WALLACH

BRONSON	BISSELL & DE HOYOS
BRYNNER	LUCERO & NAVARRO
BUCHHOLZ	BRAVO & SUAREZ
COBURN	ALONZO & VACIO
DEXTER	AVERY & RUSSELL
MCQUEEN	HERN & RUSKIN
VAUGHN	ALANIZ & WALLACH

In order to be able to work out the companions for MCQUEEN and DEXTER, knowing who accompanied BRONSON, the first of these must be the correct solution.

22   BAGS OF SWEETS!

Answer: £2-16

If the number of bags I bought is B and the price of each bag is P pence, then the total spent is T = BP

I could have bought 2 more bags at 9p less each, so (B+2)(P-9) = BP

This becomes 2P = 9B+18 = 9 (B+2), so B is even.

I could have bought a whole number of bags at 12p less per bag, so T/(P-12) is whole.

Now, T = BP = 9B(B+2)/2 and T/(P-12) = 9B(B+2)/2[9(B+2)/2-12]

This can be rewritten as B + 8/(3-2/B)

For B=2, this is 6, for B=4 it isn't a whole number and for B=6, it is 9. Larger values of B give a number less than B+3 but more than B+8/3, so not a whole number.

The total spend for B=2 is 36p and for B=6 is £2-16. Since I spent more than 50p, the latter is the required answer.

23 ALAN AND CAT

Answer: 6 March

The easiest way to count the possible routes is to work backwards from Alan's house. At each intersection, Cat can only go east or north (for the shortest possible distance) and can never end up northwest of the direct line to Alan's house (as the Street number cannot be less than the Avenue number). The number of routes from a particular intersection is the sum of those from the intersection to the north and those from the intersection to the east.

There is only one possible route from the intersection due south of Alan's house and only one from the intersection west of this. You can then build up the number of routes on the grid as follows:

								Alan's House
							1	1
						2	2	1
					5	5	3	1
				14	14	9	4	1
			42	42	28	14	5	1
		132	132	90	48	20	6	1
	429	429	297	165	75	27	7	1
1430	1430	1001	572	275	110	35	8	1

For example, if Alan lives at the junction of 5th Street and 5th Avenue, there are 14 possible routes. As Cat used a different route each day in 2018, the number of routes is over 365, but since she ran out of routes in 2019, the number must be 429. She ran out of routes after her walk on 5 March 2019 (365 routes in 2018, 31 in January 2019, 28 in February 2019 and 5 in March 2019), so she repeated a route on 6 March.

Mathematical note: the numbers of possible routes (1, 2, 5, 14, 42, 132, 429, 1430, etc.) are known as Catalan numbers. There is a formula which gives the nth value: $(2n)!/[n!(n+1)!]$

24   BOOK RECEIPTS

Answer £7.84

Let a) to i) below represent each of the cost digits. As each of the digits 1 – 9 has been used its total of 45 has a digital root of 9. It follows that both sides of the equation, including the Total Cost has a digital root of 9.

To find the maximum price that one of the books could have cost the Total Cost needs to be as high as possible and the Lower Cost Book as small as possible. So starting with a) being 1, d) being 8, and g) being 9 there are six possible Total Costs as shown below. There are only six as the Total Cost has to be divisible by 9 and can't include the digits 1 and 8 already used.

A quick look at the available numbers for e), f), b) and c) shows that none of these six possible total costs is a solution. For example for the Possible Total Cost 972 no combination of the four available numbers for c) and f) will add up to the 2 shown for i).

The next logical combination to try is changing d) from 8 to 7. This gives four possible total costs of which two produce solutions. The biggest of these solutions for d), e) and f) of 784 is in fact the solution (£1-52 + £7-84 = £9-36). This can be checked against other likely combinations as shown below which produce smaller figures.

	Lower Cost Book			+	Higher Cost Book			=	Total Cost			Possible Total Costs	Available numbers for e,f,b,c			
	a	b	c	+	d	e	f	=	g	h	i					
Try	1				8				9							
												(972	3	4	5	6
												(963	2	4	5	7
										No Solutions		(954	2	3	6	7
												(945	2	3	6	7
												(936	2	4	5	7
												(927	3	4	5	6
	1				7				9							
										No Solutions		(963	2	4	5	8
												(954	2	3	6	8
		6 or 8	2 or 3		8 or 6	3 or 2						945	2	3	6	8
		5 or 8	2 or 4		8 or 5	4 or 2						936	2	4	5	8
	2				7				9							
		3 or 4	5 or 6		4 or 3	6 or 5						981	3	4	5	6
		1 or 4	5 or 8		4 or 1	8 or 5						963	1	4	5	8
		1 or 3	6 or 8		3 or 1	8 or 6						954	1	3	6	8
										No Solutions		(945	1	3	6	8
												(936	1	4	5	8
	1				7				8							
		2 or 3	9 or 5		3 or 2	5 or 9						864	2	3	5	9
										No Solution		846	2	3	5	9

25 A RESULT

Answer: 5, 6 and 9

Using [..] to indicate either way round, possible results for 2 digit numbers are:

Result = 1 from [13], [19], [23], [28], 44, [49], [68], 70 and [79].

Result = 2 from 11 and [78].

Result = 5 from [12].

Result = 8 from 22.

Result = 9 from 30.

Result = 4 from the remaining pairs.

So one of the digits being 5 will give a result of 4.

The key digit = 5, with a result = 4.

The possible 2 digit pairs to which I add a 3rd digit are [15], [25], [35], [45], 50, 55, [65], [75], [85] and [95].

This 3rd digit must give a result of 5. The only possibility is [569].

{Result = 1 comes from [356] and [556].

Result = 2 comes from [257], [259], [559] and [567].

Result = 5 comes from [569].

Result = 9 comes from [152], [459] and [568].

Result = 4 from the remaining 3 digit numbers.}

26 'BINGO A GO-GO' LINGO A NO-GO

Answer: 1, 4, 9, 25, 36, 49, 64

If the prime factorization of T=[P^j][Q^k][R^m][S^n]... for P, Q, R, S... different primes and j, k, m, n... integers>0, then T has F=[j+1][k+1][m+1][n+1].. total factors (incl. 1 and T)

From this, no number<100 has F>=13 (F=13 T=P^12, F=14=2x7 T=P^13 or PQ^6, etc. all have prime factorization forms giving T>99). So we look at the forms for T for F=1 to F=12

F	valid prime factorizations	T values possible
1	P^0	1
2	P	all primes<100
3	P^2	4, 9, 25, 49
4	P^3; PQ	8, 27, 2x3...2x47, 3x5...3x31, 5x7...5x23, 7x11, 7x13
5	P^4	2^4=16, 3^4=81
6	P^5; PQ^2	32, 4x3...4x23, 9x2, 9x5...9x11, 25x2, 25x3
7	P^6	2^6=64
8	PQ^3; PQR	8x3...8x11, 27x2, 2x3x5...2x3x13, 2x5x7
9	[P^2][Q^2]	4x9=36
10	PQ^4	3x16=48, 5x16=80
11	none P^10>100	none
12	PQ^5; [P^2][Q^3]; PQR^2	32x3=96, 9x8=72, 3x5x4=60, 3x7x4=84, 2x5x9=90

In this teaser any set above having more than 9 members would lead to ambiguity and preclude a solution with a certain win after just nine calls. This rules out F=2 (the primes); F=4; F=6; and F=8.

Find winning combinations from valid remaining sets

F	Numbers
1	1
3	4, 9, 25, 49
5	16, 81
7	64
9	36
10	48, 80
12	60, 72, 84, 90, 96

No. of different F sets
Factor values

2	60, 72, 84, 90, 96	4, 9, 35, 49			
	Can calculate all with certainty				
3	60, 72, 84, 90, 96	48, 80	16, 81		
	Can calculate all with certainty				
4	60, 72, 84, 90, 96	48, 80	with	[1, 36] or [1, 64] or [36, 64]	
4	60, 72, 84, 90, 96	16, 81	with	[1, 36] or [1, 64] or [36, 64]	
4	4, 9, 25, 49	48, 80	16, 81	with 1 or 36 or 64	
	Can calculate none with certainty				
5	4, 9, 25, 49	48, 80	1	36	64
5	4, 9, 25, 49	16, 81	1	36	64
	Can calculate 4, 9, 25, 49, 1, 36, 64 with certainty				

Answer: 56 and 80

The winners from the groups form a knockout; so there are 2^M groups (each of K teams).

So the total number of games played last year was

$$\tfrac{1}{2}\,K(K-1)2^M + 2^M - 1$$

$$\underbrace{\qquad\qquad}_{\text{in groups}}\;\underbrace{\qquad}_{\text{in knockout}}$$

This year, similarly, there were 2^{M+N} groups each of L teams. Since the same number of games were played we have

$$\tfrac{1}{2}\,K(K-1)2^M + 2^M - 1 = \tfrac{1}{2}\,L(L-1)2^{M+N} + 2^{M+N} - 1$$

This tidies up to

$$K^2 - K + 2 = 2^N(L^2 - L + 2) \quad (*)$$

Last year's number of players was $K.2^M$ (<100) where M is at least 3 because there were quarter-finals. Hence K is at most 12. Also K cannot be 1, 2, 4 or 8 for otherwise the number of teams entering would be a power of 2 and so we *could* have a straightforward knockout competition. So K is one of the following values:

K	3	5	6	7	9	10	11	12
$K^2 - K + 2$	8	22	32	44	74	92	112	134

Now the table for L (<K) will be similar but we need $K^2 - K + 2$ to be 2^N times $L^2 - L + 2$. The only places where this happens are:

L	K	N From(*)	No of teams last year	No of teams this year	comment
3	6	2	6.2^M	3.2^{M+2}	No of teams doubled this year – inadmissible
5	7	1	7.2^M	5.2^{M+1}	**O.K.**

Finally, M is at least 3, and 7.2^M is a two-figure number. So M=3, there were 56 teams last year and 80 this year.

Answer: 16% (or 1.16 times) faster

When Sarah and Jenny first meet, their two paths complete a circumference of the track, so their combined angle travelled is 360^0. If Jenny's angular speed is J, then Sarah's is $J(1+n/100)$, where n is the percentage by which Sarah is faster (and n is 20 or less). Their combined speed is $J(2+n/100)$, so the time taken to meet is $36000/[J(200+n)]$ and Jenny has travelled $36000/[200+n]$ degrees.

The same applies for each pass, so after 7 passes, Jenny has travelled $36000*7/[200+n]$ degrees, which must be a whole number. The only value for n up to 20 which works is 10, and Jenny has travelled 1200^0, so they meet 120^0 anticlockwise (or 240^0 clockwise) from the start.

After Sarah changes her speed, they meet at the exit after a "few passes". If the number of passes is p, then Jenny has travelled $36000p/[200+n]$ degrees, where n is the new percentage by which Sarah is faster (so is not 10). By looking at the factors of the numbers from 201 to 220, we can work out the smallest possible number of passes for which Jenny has travelled a whole number of degrees.

n	factors of 200+n	Smallest value of p
1	3, 67	67
2	2, 101	101
3	7, 29	203
4	2, 2, 3, 17	17
5	5, 41	41
6	2, 103	103
7	3, 3, 23	23
8	2, 2, 2, 2, 13	13
9	11, 19	209
10	not allowed	
11	211	211
12	2, 2, 53	53
13	3, 71	71
14	2, 107	107
15	5, 43	43
16	2, 2, 2, 3, 3, 3	3
17	7, 31	217
18	2, 109	109
19	3, 73	73
20	2, 2, 5, 11	11

The only value of n which can allow a "few passes" is 16.

Answer: 9744

The maximum values of the 'e e' of the sum are 8 and 8. 188 times 8 gives 1504, but the 'thousands' digit of the first line of the multiplication is even, so the first digit is more than '1'. That digit times the 'tens' digit of the multiplier is less than or equal to 8, so we have

$$
\begin{array}{r}
3\,e\,e \\
\times\ \ 2\,e \\
\end{array}
$$

Look for values of 3 e e which give e o e when multiplied by 2, so 3 e e must be less than 350. Trial and error gives 306, 308, 326, 328, 346, 348.

Trial and error again shows that none of these multiplied by 4 or by 6 gives e o e e as the first line of the multiplication, so the multiplier must be 8 and the second number is 28.

This reasoning eventually leads to

$$
\begin{array}{r}
348 \\
\times\ \ 28 \\
\hline
2784 \\
696\ \ \\
\hline
9744 \\
\end{array}
$$

Answer: 99

To start with, we need to list the candidate two-digit numbers (11- 31 inclusive as any more squares to over 1000) and their three-digit perfect squares. Anything ending in 1 5 or 6 is out automatically so we are left with the following:

12	13	14	17	18	19	23	24	27	28	29
144	169	196	289	324	361	529	576	729	784	841

The only cases where five different digits appear are

17	18	24	29
289	324	576	841

We now list single-digit candidates and their squares – the same rules as above applying except that we have to start at 4 to get a two-digit square:

4	7	8	9
16	49	64	81

The only combination in which eight different digits appear is 24/576 and 9/81.

Thus the garden is a 24 by 24 square for 576 and the flower-bed is a 9 by 9 square for 81 leaving the surrounding lawn an area 495. So 3 is still missing and therefore the gardener works for three hours i.e. 180 minutes.

Suppose s = number of minutes worked on the surrounding lawn; then $180 - s$ = number of minutes worked on the flower bed. Thus we have:

$$ns + 180 - s = 576 \text{ and thus } (n - 1)s = 396$$

Thus we need to factorise 396 to get candidates for $n - 1$ and s, bearing in mind that n must be a single digit.

$n-1$	s	
2	198	out – must be less than 180
3	132	possible: $n = 4$ but $f = 148$ not a square
4	99	possible: $n = 5$, $f = 81 = 9$ squared as required
6	69	possible: $n = 7$, $f = 111$ not a square

Thus $n = 5$ and $s = 99$.

Answer: 545/252

15	14	13	12	11	10	9	8	7	6	5	4	3	2	1
1	2	3	4	5	6	7	8	9	10	11	12	13	14	15

The selection from the 15 'ruler overlap' sets uses only numerals 1, 2, 3, 4, 5, 6, 7, 8 and 9 – each once only. Only a selection of four fractions works, involving one 2-digit value as a denominator (from 10, 11, 12, 13, 14 or 15) and seven single-figure values.

All are fractions with values less than one (**numerator<denominator**).

2-fig. denominator (1n) can't be 10 or 11 and must be 12, 13 or 14, as only these can be part of a fraction not repeating a numeral (15 only occurs in 1/15 – invalid) i.e. **2/14; 2/13; 3/12 or 4/12** from the fifteen sets (from 1/15, 1/14, **2/14**, 1/13, **2/13**, 3/13, 1/12, 2/12, **3/12**, **4/12**).

Two of the four fractions are in simplest form (one with consecutive nom/den), so two are reducible.

In general (prior to simplification, if possible)
[P/Q+R/S+T/U+V/W]=[P*S*U*W+R*Q*U*W+T*Q*S*W+V*Q*S*U]/[Q*S*U*W]

With 2/14 or 4/12 (both reducible), we need three proper fractions using 3, 5, 6, 7, 8 and 9 – with just one reducible. Only 3/6, 3/9, 6/9 and 6/8 are reducible (3 must be a numerator; 9 must be denominator; and 8/9 is not in the fifteen sets, so 8 must also be a denominator and simplest form consecutive nom/den=5/6 or 6/7 or 7/8)

So possible tetrad sums are, from above and in simplest form:-

[2/14+3/6+7/8+5/9]=12540/6048=**1045/504**; [2/14+3/9+5/8+6/7]=13818/7056=**47/24**
[2/14+6/9+3/5+7/8]=11514/5040=**1919/840**

[4/12+3/6+7/8+5/9]=11736/5184=**163/72**;   [4/12+3/9+5/8+6/7]=12996/6048=**361/168**
[4/12+6/9+3/5+7/8]=10692/4320=**99/40**

With 2/13 (simplest form) we need three proper fractions (two reducible) from 4, 5, 6, 7, 8 and 9. Reducible options are 4/6, 4/8, 6/8 and 6/9 – only possible pairing is 4/8 with 6/9, so only tetrad is [2/13+4/8+6/9+5/7] – invalid – 2/13 and 5/7 not consecutive nom/den

With 3/12 (reducible) we need three proper fractions (one reducible) from 4, 5, 6, 7, 8 and 9.

Reducible options are 4/6, 4/8, 6/8 or 6/9 with one of 4/5, 5/6, 6/7 or 7/8; so valid tetrads are:-

[3/12+4/6+5/9+7/8]=12168/5184=**169/72**;   [3/12+4/8+5/6+7/9]=12240/5184=**85/36**
[3/12+4/8+5/9+6/7]=13080/6048=**545/252**;   [3/12+6/8+4/5+7/9]=11136/4320=**116/45**

Only answer with palindromic numerator and denominator is from
3/12+4/8+5/9+6/7=[3x7x8x9+4x7x9x12+5x7x8x12+6x8x9x12]/[7x8x9x12]= 13080/6048=[24x545]/[24x252]=545/252

Answer: 26198073

Let the original number (N) be a square.
It must have cubes as its digit number and digit sum.
These could only be 8 and 27.
The sum of eight different digits cannot be less than 28 so the number is not a square.

N is a cube and must have 4, 8 or 9 digits.
If it is 4 or 9 digits the sum of the digits must be 8 or 27 so 9 digits is impossible.
As 11^3 is 1331 we can rule out 4 digits.

N is a cube with 8 digits and digit sum 36 (so is divisible by 9).
N must be of the form 33^3.n^3 where n is an integer.

n	7	8	9	11	12	13	14
N	12326391	18399744	26198073*	47832147	62099137	78953589	98611128

All the digits must be different, so N is 26198073

33 SIX SISTERS ON THE SKI LIFT

Answer: 7, 10, 13, 15, 21 and 26

Because no three of the girls' ages has a common divisor, but any two girls standing one behind another do have ages with a common divisor, we can assume that there is a different prime number that divides each of the first five pairs of girls' ages in the line (that is five primes in all), while at least two primes divide each of the ages of the four middle girls.

In order to minimize the sum of the ages and not exceed 92, let us assume that the five primes involved are the smallest five primes 2, 3, 5, 7, and 11, and place 11 on one end, and 5 or 7, at the other end (any other choice for extreme primes results in sums greater than 92). In each case, checking the six ways of placing the remaining three primes in between 11 and 7 (or 11 and 5) yields only two cases in which the sum is less than 92 (and none equal to 92): 11,22,14,21,15,5 for a total of 88, and 11,22,10,15,21,7 for a total of 86.

Thus we need to use at least one prime larger than the five smallest primes, the obvious choice being 13. Here again we need to minimize the impact of the two largest primes of our choice: 13 and 7 (or 5), so we try 13 on one end, and 7 (or 5) on the other. Checking the six ways of placing the remaining three primes in between 13 and 7 (or 5), yields only one case where the sum is not greater than 92, and it is precisely 92: 13, 26, 10, 15, 21, 7.

It is straight forward that the use of any prime greater than 13 would result in sums greater than 92.

34   A RELAXING DAY

Answer: 19:56:48

First of all consider the digits 6, 7, 8 and 9, which have to appear eight times in total. They can appear up to three times in the seconds, up to three times in the minutes, so must appear at least twice in the hours. The only choice for the hours which does not include three 1s is 16, 19 and 23 (can't have 22 as the 2 is repeated).

As there is an exact number of minutes difference between the lengths of the sleeps, the last digits of the seconds must form an increasing or decreasing sequence, so they are (6,7,8), (8,7,6), (7,8,9) or (9,8,7). Similarly, the first digits of the seconds must form a sequence, so the only possibilities are (2,3,4), (4,3,2), (3,4,5) and (5,4,3).

Moving on to the minutes, the last digits will be (6,7,8) or (7,8,9) in some order, while the first digits will be (2,4,5) or (4,5,5) in some order. (4,5,5) doesn't work, since there cannot be a 23 minute difference in sleep times and for the other possibility, the order must be (4,5,2) or (2,5,4). (4,5,2) doesn't work since a 2 in the last time will repeat the 2 in 23 hours. The corresponding first digits of the seconds will be (3,4,5) (not (5,4,3) since a 3 in the last time will repeat the 3 in 23 hours). The possibilities are now:
16:2a:3d then 19:5b:4e then 23:4c:5f, where a, b and c are (6,7,8) in some order and d, e and f are (7,8,9) or (9,8,7).

To produce a 23-minute difference in sleeps, the only solutions are:
16:28:37 then 19:56:48 then 23:47:59 or
16:27:39 then 19:56:48 then 23:48:57

In either case, George wakes up after his first sleep at 19:56:48

Answer: 278 and 417

If the first two numbers are X and Y then $X = A \times HCF$ and $Y = B \times HCF$ for some A, B with no factor in common. Then $LCM = A \times B \times HCF$.

Since the four numbers are different, A and B are greater than 1, but $A \times B$ must be less than 10. The only possibility is that A and B are 2 and 3.

Therefore $LCM = 6 \times HCF$ and H=1. We tabulate the possibilities:

F	M	(6F-M)/10
3	8	1
5	0	3
7	2	4
9	**4**	**5**

Now $LCM = 6 \times HCF$ means

$100L + 10C + M = 600 + 60C + 6F$
i.e. $(6F-M)/10 = 10L - 60 - 5C$

and so $(6F-M)/10$ must be divisible by 5, as in the final case above.

Therefore F=9, M=4 and $10L - 60 - 5C = 5$. So $2L - C = 13$ and (since L $\neq$ 9 and C $\neq$ 1) we have L=8 and C=3:

$HCF = 139$, $LCM = 834$

the first two numbers are $2 \times 139 = \textbf{278}$ and $3 \times 139 = \textbf{417}$

Answer: 99

Wu; Xi; Yo and Ze flock sizes **W, X, Y** and **Z** – all different >9 and <100 with **W+Z=X+Y** and **W=(p/q)X** and **Y=(q/p)X** (**W** largest so p > q >0 are consecutive numbers so q=p-1, hence no non-trivial common factors, hence fractions in simplest form)

So **X must divide by p and q** – and for k=positive integer, **X=pqk** and hence

W=ppk (so p<10, else W>=100) and **Y=qqk** and **Z=X+Y-W=(pq+qq-pp)k**

Total sheep=T=W+X+Y+Z=2q(q+p)k=Nk (N positive integer)
and this must have 4 as a factor, because same number of sheep in each pen.

Tabulate, **with k=1**, for possible consecutive single-figure p > q pairs

p <10	q= p-1	W= ppk	X= pqk	Y= qqk	Z= (pq+qq-pp)k =X+Y-W	T= W+X+Y+Z =Nk	T/4= Nk/4
9	8	81	72	64	55	272	68
8	7	64	56	49	41	210	52.5
7	6	49	42	36	29	156	39
6	5	36	30	25	19	110	27.5
5	4	25	20	16	11	72	18
4	3	16	12	9	5	42	10.5
3	2	9	6	4	1	20	5
2	1	4	2	1	not valid<0		

9<W, X, Y, Z<100 implies the mean value = T/4 <100. Tabulate valid options for k>=1

p/q	T/4	W/X/Y/Z	k	T/4	W/X/Y/Z	k	T/4	W/X/Y/Z	k
9/8	68	81/72/64/55	1						
8/7	none valid								
7/6	39	49/42/36/29	1	78	98/84/72/58	2			
6/5	**55**	72/60/50/38	**2**						
5/4	18	25/20/16/11	1	36	50/40/32/22	2	54	75/60/48/33	3
4/3	21	32/24/18/10	2	42	64/48/36/20	4	63	96/72/54/30	6
3/2	50	90/60/40/10	10	**55**	**99/66/44/11**	11			
2/1	invalid								

Only 'ambiguous' T/4 = 55, but it can't be allied to 'unambiguous' p/q=6/5 [q/p=5/6] so it must be allied to 'ambiguous' p/q=3/2 [q/p=2/3] and so W=99

Answer: 3754

The number of bricks N required to build a hollow cube of size x is given by
$N = x^3 - (x - 2)^3 = x^3 - (x^3 - 6x^2 + 12x - 8)$
$N = 6x^2 - 12x + 8$ ....(1)

For a child let the cube sizes be A and B. Let the father's cube be size C then
$N_A + N_B = N_C + 2$

substituting for N from equation (1) gives
$(6A^2 - 12A + 8) + (6B^2 - 12B + 8) = (6C^2 - 12C + 8) + 2$

which simplifies to
$A^2 + B^2 = C^2 + 2(A + B - C) - 1$ ....(2)

A, B and C are integers so the solutions to equation (2) will be related to the integer solutions for Pythagoras's right-angled triangles $a^2 + b^2 = c^2$. The first integer solution for Pythagoras, is the triple 3, 4, 5 and the first solution to equation (2) is A = 4, B = 5, C = 6.

Substituting A = (a + 1), B = (b + 1) and C = (c + 1) into equation (2) gives
$(a + 1)^2 + (b + 1)^2 = (c + 1)^2 + 2((a + 1) + (b + 1) - (c + 1)) - 1$ .....expanding this gives
$(a^2 + 2a + 1) + (b^2 + 2b + 1) = (c^2 + 2c + 1) + 2(a + b - c + 1) - 1$ .....which simplifies to
$a^2 + b^2 = c^2$

So for each integer solution for $a^2 + b^2 = c^2$ there is an A= (a + 1), B= (b + 1), C= (c + 1) solution to equation (2).

The cube problem can now be resolved by finding two different right-angled triangles having integer sides and the same hypotenuse c, where c is the smallest possible value.

The first four primitive Pythagorean triples (triples which cannot be constructed from multiples of smaller triples) are (3, 4, 5), (5, 12, 13), (8, 15, 17) and (7, 24, 25). The first three of these have prime number values 5, 13 and 17 for c, so they cannot be matched with multiples of smaller triples. However, the triple (7, 24, 25) has c = 25 and can be coupled with (15, 20, 25), a multiple of 3, 4, 5. Therefore 25 is the smallest value of c to appear in two Pythagorean triples.

As c = 25 then the father's cube C must be 26. Adding one to the values for a and b in the two triples gives (8, 25) and (16, 21) for each of the children's hollow cube sizes.

Using equation (1) the number of bricks in the father's cube is
$(6 \times 26 \times 26) - (12 \times 26) + 8 = 4056 - 312 + 8 = 3752$
and as the father has two bricks left, we must add two more, so

The number of bricks in the box is 3754

Checks and validation.
The fathers cube contains $(26^3 - 24^3)$ bricks and adding 2 = 17576 − 13824 + 2 = **3754**
1st child's cubes $(8^3 - 6^3) + (25^3 - 23^3) = 512 - 216 + 15625 - 12167 = $ **3754**
2nd child's cubes $(16^3 - 14^3) + (21^3 - 19^3) = 4096 - 2744 + 9261 - 6859 = $ **3754**

Answer: 54 ounces

Top/base are regular octagons (all vertical faces equal squares, so all sides equal length L; with height H=L=AB, giving (see diagrams below)

AF=octagon span=(1+2/√2)AB=(1+√2)AB

Base area, A=(AFxAF)-(ABxAB)=(2+2√2)(L^2)=2(1+√2)(L^2)

(large square–small square made of 4 corner pieces) and with vertical faces making a prism

Volume of cake, V=AxH=AxL=2(1+√2)(L^3)

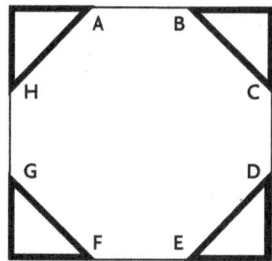

With 8<=N=no. present<=12 (me + majority of 12 invited guests, not all).

Portion size in ounces=Portion volume in cu. in.=P=integer part of V/N

Leftover portion=R=V-NxP

P is 2-figure integer, so 3<=L<=6, as P<10 for L=2, N>=8 and P>99 for L=7, N<=12

Tabulate with 3<=L<=6, 7=(8-1)<=N<=13=(12+1) showing P with [integer part of R]

L (in.)	3	4	5	6
V (cu. in.)	130.3..	309.02..	603.5..	1042.9..
N				
7	18 [4]	44 [1]	86 [1]	>99
8	16 [2]	38 [5]	75 [3]	>99
9	14 [4]	34 [3]	67 [0]	>99
10	13 [0]	30 [9]	60 [3]	>99
11	11 [9]	28 [1]	54 [9]	94 [8]
12	10 [10]	25 [9]	50 [3]	86 [10]
13	10 [0]	23 [10]	46 [5]	80 [2]

L=3, N=7, {8}, 9 give same increased residual portion – invalid.

L=4, N=10, {11}, 12 give same increased residual portion – invalid.

L=5, N=8, {9}, 10 give same increased residual portion – invalid.

L=5, N=10, {11}, 12 give same reduced residual portion.

So my portion was 54 ounces.

Answer: 8563

We need to look at combinations of digits in which the sum, difference, product and average are all unequal. Clearly therefore 1 is ruled out and we know that, as the average was a digit, the two digits must be both even or both odd to have any chance at all. Furthermore, 5 times any odd digit will result in 5 at the end so that too is out. 6 multiplied by any even digit leads to that digit appearing in the product so that too is out. Any two digits adding to ten are also ruled out and there must be no repetitions. We are left with the following four possibilities:

Last digit of Martha's	3	9	9
Last digit of George's	9	3	7
Sum	2	2	6
Difference	4	6	2
Product	7	7	3
Average	6 or 1	1 (6 out)	8 (3 out)

In the first column a b c are 1 5 8 or 5 6 8
In the second, a b c are 4 5 8
In the third a b c are 1 4 5

We now consider the averages. In the first case, we have xx66 averaging abc3 and the smaller abc9 where the abc's 1, 5, and 8 are interchangeable.

Consider combinations of the two a's: 1 and 1 implies 1166
       1 and 5 implies 3366
       1 and 8 implies 4466 or 5566
       5 and 5 implies 5566
       5 and 8 implies 7766
       8 and 8 implies 8866

None of these are possible on our rules.

In the second case, we have xx11 averaging abc3 and the smaller abc9 where the abc's 5, 6 and 8 are interchangeable.

Following the line above: 5 and 5 implies 5511
       5 and 6 implies 6611
       5 and 8 implies 6611 or 7711
       6 and 6 implies 6611
       6 and 8 implies 7711
       8 and 8 implies 8811

The only one which works is 8563 and 6859 averaging 7711

Working similarly, there are no other possibilities on the two arrangements so the answer is 8563.

Answer: 243000 sq metres

Very little trial and error drawing shows that we are dealing with a regular hexagon, where each son's areas is in the shape of a trapezium.
The triangle is formed by joining alternate midpoints.
The easiest solution is via further construction lines.
Draw another triangle by joining the other midpoints.
Draw in the three diagonals of the hexagon.
We now have 24 congruent triangles; 5 for each son and 9 for Sam.

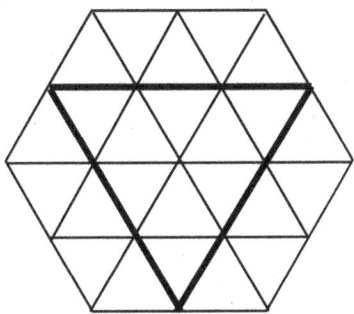

Let Sam's area be A and that for each of the sons be B, then $5A=9B$.

There are two cases:

(a) Sam has the cube, each son the square

For A to be a cube, B must be $15 \, m^3$ where m is a whole number, so $A = (3m)^3$.
For B to be a square, m must be $15 \, n^2$ where n is a whole number, so $B = (225 \, n^3)^2$
Then $A = 91125 \, n^6$ and $B = 50625 \, n^6$. The total area is $A+3B = 243000 \, n^6$.
The only possibility is n=1, otherwise the area is more than 100 hectares.

(b) Sam has the square, each son the cube

For B to be a cube, A must be $225 \, p^3$, where p is a whole number, so $B = (5p)^3$
For A to be a square, p must be a square (q^2), so $A = (15 \, q^3)^2$.
Then, $A = 225 \, q^6$ and $B = 125 \, q^6$. The total area is $A+3B = 600 \, q^6$.
q = 1 gives an area of 600 sq metres. q = 2 and 3 give areas 64 and 729 times larger.
All three cases are less than 100 hectares.

In case (a), we can work out the area of Norfolk Flats, but in case (b) we can't. Therefore, case (a) is the correct one and Norfolk Flats has an area of 243000 square metres.

41   MISCHIEVOUS SAM

Answer: (a) 123 and (b) 8

* The base needs to be at least 5 in order to achieve a 3-digit number
* NOTE. A neat way to change a number to a different base is to list the remainders when dividing by that base.
e.g. to change 123 to base 9: 123/9 = 13, remainder 6 and 13/9 = 1, remainder 4 so 123 = 146, base 9

(in the following table, – means 4 digits or more)

Base	123	234	345	456	567	678	789
5	443	–	–	–	–	–	–
6	323#	–	–	–	–	–	–
7	234	453	–	–	–	–	–
8	173#	352	531	710	–	–	–
9	146	280	423	556#	700	833	–

The table shows that 323 and 556 have the 1st digit wrong and 173 has the 2nd digit wrong.

And so the original answer was 123 and this was changed to base 8 to give 173.

Answer: 2, 5, 8, 10, 25, 40, 50, 100, 250, 500g

The possible weights that are unit fractions of a kilo are as follows (in grams):

1, 2, 4, 5, 8, 10, 20, 25, 40, 50, 100, 125, 200, 250, 500

It is possible (as we will demonstrate) to use a selection of weights totalling just 990g to weigh any amount in 10g steps from 10g to 990g, giving the minimum amount of brass.

If we use a 1g weight, it must be in conjunction with the 4g. Similarly, the 2g must go with the 8g. It is easier to regard these as additional 5g and 10g weights (but remembering that they consist of two weights).

We need weights totalling 990g, which has a remainder of 15 on division by 25.
Now consider the remainders on division by 25 of all the possible weights:

Wt	5	5	10	10	20	25	40	50	100	125	200	250	500
Rem	5	5	10	10	20	0	15	0	0	0	0	0	0

It is clear that 10g (or 5g+5g) and 500g will always be needed, together with one of 200g and 250g.

If 40g is not included, then the remainders must total 15 (5+10) or 40 (5+5+10+20 or 10+10+20). If 15, then the weights are 5, 10, 25... so it is impossible to weigh 20g.

Therefore the remainders total 40. The four possibilities for weights adding to 990g are as follows:

5, 5, 10, 20, 25, 50, 125, 250, 500 (cannot weigh 120g since 5+5+10+20+25+50=115)

5, 5, 10, 20, 25, 100, 125, 200, 500 (cannot weigh 70g since 5+5+10+20+25=65)

10, 10, 20, 25, 50, 125, 250, 500 (cannot weigh 120g)

10, 10, 20, 25, 100, 125, 200, 500 (cannot weigh 70g)

Therefore the 40g weight is needed. The other remainders must total 25 (5+10+10) or 50 (5+5+10+10+20). There are four possibilities for weights adding to 990g:

5, 10, 10, 25, 40, 50, 100, 250, 500 ✓

5, 10, 10, 40, 100, 125, 200, 500 (cannot weigh 30g)

5, 5, 10, 10, 20, 40, 50, 100, 250, 500 ✓

5, 5, 10, 10, 20, 25, 40, 50, 125, 200, 500 ✓

The first possibility gives the solution with the fewest weights. The full list of weights is:

2, 5, 8, 10, 25, 40, 50, 100, 250, 500

Answer: 10080

Let X be the true number of steps and Y the pedometer count of steps, then $Y = 37225$ and $X = 0.68*37225 = 25313$.

Let the rollover values for the five digits be a, b, c, d, e. In other words, the final digit rolls over to 0 instead of to e, so if $e = 7$ then the last digit rolls over from 6 to 0. This is just like counting in base e rather than base 10. The number of steps from the last two digits (25) will be $2e + 5$, since each time the last digit goes to 0 there will have been e steps. Carrying on in this way, we see that:
$X = 3bcde + 7cde + 2de + 2e + 5 = 25313$

Then $25308/e$ is a whole number, but e must be greater than 5 (otherwise the final digit cannot display a 5). The possible only values are $e = 6$ and $e = 9$, giving $3bcd + 7cd + 2d = 4218$ or 2812. The first of these is impossible, since the largest possible values of b, c and d are 10

For $e = 9$, $3bcd + 7cd + 2d + 2 = 2812$, so $2810/d$ is a whole number, and $d = 5$ or 10, giving $3bc + 7c + 2 = 562$ or 281. The first is impossible, so $3bc + 7c = 279$ and the only possibility is $c = 9$ and $b = 8$.

For the second distance, let $Y = ABCDE$. Using the values b=8, c=9, d=10 and e=9, we have:
$X = 6480A + 810B + 90C + 9D + E$.

We know that this is 70% of the displayed value, so
$6480A + 810B + 90C + 9D + E = 0.7(10000A + 1000B + 100C + 10D + E)$.

A to E are single digits, so $E = 0$. The equation becomes:
$520A = 110B + 20C + 2D$

D can only be 0 or 5 and A can be no more than 2. The possible solutions are:

A	B	C	D	E	
1	4	4	0	0	
1	3	9	5	0	No, since C >= c
2	8	8	0	0	No, since B >= b
2	9	2	5	0	No, since B >= b

Three solutions are ruled out because a digit exceeds the relevant rollover value.

Therefore, $Y = 14400$ and $X = 10080$.

Answer: £350

Look at the final Ns ∴ N = 5 or 6

Look at P x N = P ∴ N = 6 only

Look now at M A 6 multiplied by I to give A A S A

Either I = 3, A = 8 ∴ S = 5 ∴ M x

Or I = 9, A = 4 ∴ S = 1 ∴ M x

Or I = 7, A = 2 ∴ S = 8 ∴ M = 3 ∴ O = 5 ∴ U = 9 ∴ T = 1 ∴ P = 4 ∴ E = 0.

So his code is:

0	1	2	3	4	5	6	7	8	9
E	T	A	M	P	O	N	I	S	U

∴ M O E = £350.

Answer: 147

Looking at the sum, the lowest theoretical total would put the four highest digits into the units line, typically $10 + 26 + 37 + 48 + 59 = 180$. At the other extreme, we put the lowest digits into the units line, typically $50 + 61 + 72 + 83 + 94 = 360$. So our total lies between those two. It must be divisible by 9 (since any sum of numbers comprising the digits from 0 to 9 must be divisible by 9), so the candidates are:
180 189 198 207 216 225 234 243 252 261 270 279 288 297 306 315 324 333 342 351 360, and if we divide each by 9 we get 20 – 40 inclusive.

We need to find a sum which has factors including three two-digit numbers with all different digits. We can eliminate the multiples of 11 (repeated digits) and prime numbers (insufficient factors) immediately, leaving:
20 21 24 25 26 27 28 30 32 34 35 36 38 39 40

We can now eliminate most of the multiples of 3 (extra multiple of 9), viz.

21: $3 \times 9 = 27$　$7 \times 9 = 63$
24: $2 \times 9 = 18$　$4 \times 9 = 36$　$8 \times 9 = 72$
27: cannot avoid multiples of 9
36: 2×3^4 same applies
39: gives either 13 or 39 but not both; same applies

We are left with 20 25 26 28 30 32 34 35 38 40

$20 \times 9 = 180$ boils down to $5 \times 3 \times 3 \times 2 \times 2$; the only two possible factors not ending in 0 are 18 and 36, but both are divisible by 9

$25 \times 9 = 225$ boils down to $5 \times 5 \times 3 \times 3$: cannot have five divisors without repeating digits

$26 \times 9 = 234$ boils down to $2 \times 3 \times 3 \times 13$, allows 18 26 39 78, leaving 4 5 0; no good

$28 \times 9 = 252$ boils down to $2 \times 2 \times 3 \times 7 \times 9$ allows 12 14 18 21 27 28 36 42 54 63 84, gives possible solution of 14 28 63 leaving 50 and 97 to add to 252 as required

$30 \times 9 = 180$ boils down to $5 \times 3 \times 3 \times 3 \times 2$; the only way to get two factors not divisible by 9 is to have 15, 30 and 27. This leaves 4, 6, 8 and 9 and we can't get a sum ending in 0.

$32 \times 9 = 288$ boils down to $2^5 \times 3^2$ insufficient possibilities

$34 \times 9 = 306$ boils down to $2 \times 9 \times 17$ allows 17 18 34 51 insufficient possibilities

$35 \times 9 = 315$ boils down to $3 \times 3 \times 5 \times 7$ allows 15 21 35 45 63 insufficient possibilities

$38 \times 9 = 342$ boils down to $2 \times 9 \times 19$ allows 18 38 57 insufficient possibilities

$40 \times 9 = 360$ The smallest possible number is 50, so the factors are 60, 72 and 90 – no good

46 SHUFFLE THE CARDS

Answer: 130

It is immediate from the story-line that none of the three chosen digits was zero.

If A, B and C are different non-zero digits then A.BC cannot equal A.CB. Nor can A.BC equal C.BA (for that would require A=C) and nor can A.BC equal B.AC (for that would require A=B). So either A.BC = B.CA or A.BC = C.AB, but these are actually equivalent to each other (try substituting a for C, b for A and c for B in the second).

So we need A.BC = B.CA with A, B, C different, non-zero and with A≠1 and B≠1 (the result has three digits).

Case (a): A=5

BC = B.CA gives 50B +5C = 10B.C + 5B or B = C/(2C - 9). The only possibility (with C≠ 5 and B≠1) is C=6, B=2. This gives

$$5.26 = 2.65 = 130$$

which fits the story-line.

Case (b): A ≠ 5

C ends in the same digit as A.B. This can only work if A=5 (covered in case (a)) or if A is even and B/C differ by 5. Furthermore, that final digit of A.B is different from A, B and C. We list the possibilities:

A	B	C	A.BC	B.CA
2	3	8	76x	
	8	3	166x	
	4	9	98x	
	9	4	188x	
4	2	7	108	148x
	7	2	288x	
	3	8	152	252x
	8	3	332x	
8	2	7	216x	
	7	2	576x	
	4	9	392x	
	9	4	752	432x

So the answer found in case (a) is indeed the only one.

47 WHAT'S MY (LAND) LINE?

Answer: 0 1 8 6 7 2 9 5 3 4 0

Pairs of consecutive numbers that fit the facts are 3-4 and 6-7.

So   0 1 . 3 4 . . . 6 7 0

        0 1 . 3 4 . . . 7 6 0

        0 1 . 4 3 . . . 6 7 0

        0 1 . 4 3 . . . 7 6 0

        0 1 . 6 7 . . . 3 4 0

        0 1 . 6 7 . . . 4 3 0

        0 1 . 7 6 . . . 3 4 0

        0 1 . 7 6 . . . 4 3 0

Of the remaining digits, we cannot have 5.8.9 (or 2.5.8 or 2.8.9) at positions 6 7 8 in any order, so we must have 2 9 5 or 5 9 2. They must be 2 9 5, with the 9th and 10th digits being 3 or 4, otherwise the 8th digit is lower than the 9th. Thus, the 3rd digit is 8.

Answer: 106:01

If the prime factorization of T=[P^j][Q^k]... for P, Q,... different primes and j, k,... integers>0, then T has F=[j+1][k+1]... factors (incl. 1 and T). For T=2 to 20 we find:

T=	2	3	4	5	6	7	8	9	10	11	12
Fact.	2^1	3^1	2x2	5^1	2x3	7^1	2^3	3^2	2x5	11^1	3x2^2
Form	P^1	P^1	P^2	P^1	PQ	P^1	P^3	P^2	PQ	P^1	PQ^2
F=	2	2	3	2	4	2	4	3	4	2	6

T=	13	14	15	16	17	18	19	20
Fact.	13^1	2x7	3x5	2^4	17^1	2x3^2	19^1	5x2^2
Form	P^1	PQ	PQ	P^4	P^1	PQ^2	P^1	PQ^2
F=	2	4	4	5	2	6	2	6

For the eight prime T values the 'duration' value **130<D<960 (3-fig. m:ss format)** for each must be prime, hence odd, and – if possible – palindromic. 3-fig. palindromic prime D values are of form 1n1, 3n3, 7n7 or 9n9 with n<6 as follows:-

131, 141=3x47, **151**;
303=3x101, **313**, 323=17x19, 333=3x3x37, 343=7x7x7, **353**;
707=7x101, 717=3x239, **727**, 737=11x67, 747=3x3x83, **757**;
909=3x3x101, **919**, **929**, 939=3x313, 949=13x73, 959=7x137

So exactly eight (in **bold**) to perm with eight track numbers T=**2, 3, 5, 7, 11, 13, 17, 19**.

For T=4 and 9, D must be the square of a prime. Only D=23^2=**529** and 29^2=**841** work.

For T=16, D must be the 4th power of a prime. Only D=5^4=**625** works.

For T=12, 18 and 20, of form PQ^2, possible even palindromic D values might arise from D=2Q^2 or Px2^2=4P (but not the other 6 factors option P^5; 3^5=243 only valid value).

For D=2Q^2 the only valid options are with Q=11, 13 and 19 i.e. D=**242**, 338, 722.
For D=4P only need to consider 2n2, 4n4, 6n6, 8n8 with n2, n4, n6, n8 div by 4 for n<6:-
212=4x53, 232=8x29, 252=4x7x9; and **404=4x101**, 424=8x53, 444=4x3x37; and 616=8x7x11, 636=4x3x53, 656=16x41; and 808=8x101, 828=4x23x9, 848=16x53.

So exactly three even palindromic D values (in **bold**) to perm with **T=12, 18, 20**.

For T=6, 8, 10, 14 and 15, of form P^3 or PQ, possible even palindromic D values might arise from D=2Q (Q prime>2) and might be 2n2, 4n4, 6n6 or 8n8 with n<6 as follows:-

202=2x101, 222=2x3x37, 242=2x11x11 (see above for 212, 232 and 252)
414=2x23x9, 434=2x7x31, **454=2x227** (see above for 404, 424 and 444)
606=2x3x101, **626=2x313**, 646=2x323=2x17x19 (see above for 616, 636 and 656)
818=2x409, **838=2x419**, 858=2x3x11x13 (see above for 808, 828 and 848)

So exactly five even palindromic D values perm with five T values
Total album duration= 1:30+1:31+1:51+2:02+2:12+2:42+3:13+3:53+4:04+ 4:54+5:29+6:25+ 6:26+7:27+7:57+8:18+8:38+8:41+9:19+9:29=106:01

Answer: (a) 72 and (b) 16

An n-sided regular polygon contains $(n - 2)$ non-overlapping triangles and so each interior angle will be $(n - 2) \times 180 /n$ degrees.

Similarly an m-sided regular polygon will have interior angles of $(m - 2) \times 200/m$ grads.

$\therefore (n - 2)180/n = (m - 2)200/m \therefore n = 18m/(20 - m)$.

Since $3 < m < 20$ and m, n and $(n - 2)180/n$ are whole numbers,

m	5	8	10	11	12	14	15	16	17	18	19
n:	6	12	18	22	27	42	54	72	102	162	342
Angle:	120	150	160	-	-	-	-	175	-	-	-

{the whole number numerical value of the interior angle}

Of the 4 possible numerical values for angles, the last digit of 5 is unique.

$\therefore$ Beth has a 72-sided table mat, with interior angles of 175 degrees and Sam has a 16-sided table mat, with interior angles of 175 grads.

Answer: 6

Let there be n boxes and the amounts in the boxes be 1, 2, 3, 4, 5 and so on (the actual amounts don't matter so long as they increase in regular amounts).

Alf's strategy

If the first box contains 1, then Alf will win 2, 3, ..., or n, i.e. the first amount greater than 1. His winnings will, on average, be $(2+3+...+n)/(n-1)$, which works out as $(n+2)/2$.
If the first box contains 2, then Alf wins 3, 4, ..., or n, i.e. the first amount greater than 2, so on average $(3+...+n)/(n-2)$, which works out as $(n+3)/2$.
If the first box contains 3, then Alf wins $(n+4)/2$ on average, and so on until
If the first box contains n-1, then Alf wins $(n+n)/2$ on average, but
If the first box contains n, then Alf wins 1, 2,..., or n-1, i.e. has to take what is in the last box, which works out as $n/2$ on average.
As each of these is equally likely, his average winnings are $[(n+2)/2 + (n+3)/2 + ... + (n+n)/2 + n/2]/n$, which works out as $(3n^2+n-2)/4n$ or $(n+1)(3n-2)/4n$

Bert's strategy

If the first two boxes contain x and y, where y is larger than x but less than n, then Bert will win any amount from y+1 to n, so on average he wins the average of the numbers greater than y.
If y=n, then Bert has to take what is in the last box, so will win the average of the numbers which exclude x and y. For small values of n, these calculations are easy to do.
For n=3, Bert must take the last amount, so wins the average of 1, 2 and 3, i.e. 2.
For n=4, Bert's average winnings for the values of x and y are (x,y,winnings):
(1,2,7/2), (1,3,4), (1,4,5/2), (2,3,4), (2,4,2), (3,4,3/2), so the average win is 35/12 or around 2.92
For n=5, the average winnings are:
(1,2,4), (1,3,9/2), (1,4,5), (1,5,3), (2,3,9/2), (2,4,5), (2,5,8/3), (3,4,5), (3,5,7/3), (4,5,2), with an average of 38/10 or 3.8
For n=6, the average winnings are:
(1,2,9/2), (1,3,5), (1,4,11/2), (1,5,6), (1,6,7/2), (2,3,5), (2,4,11/2), (2,5,6), (2,6,13/4), (3,4,11/2), (3,5,6), (3,6,3), (4,5,6), (4,6,11/4), (5,6,5/2), with an average of 14/3 or 4.67

Average winnings for the two strategies

n	Alf	Bert
3	2.33	2
4	3.125	2.92
5	3.9	3.8
6	4.67	4.67

For n=6 they both win 14/3 on average. It is possible, though difficult, to work out Bert's average winnings for any value of n. This turns out to be $(n+1)(5n-6)/6n$. Putting this equal to $(n+1)(3n-2)/4n$ confirms n=6 as the answer.

More generally, if Alf's strategy is S1, Bert's is S2 and so on, then for strategy Si with n boxes, the average winnings are $(n+1)[(2i+1)n-i(i+1)]/[2n(i+1)]$

Answer: 15 and 352800

Consider the clearance in reverse i.e. from black (certain) then pink (1 in 2) etc. before starting on the 'red then colour' routine.

Inverting the fractions we have the sequence-

$$1.2.3.4.5.6.\frac{6}{6}.\frac{7}{1}.\frac{7}{6}.\frac{8}{2}.\frac{8}{6}.\frac{9}{3}.\frac{9}{6}.\frac{10}{4}.\frac{10}{6}.\frac{11}{5}.\frac{11}{6}.\frac{12}{6}.\frac{12}{6}.\frac{13}{7}.\frac{13}{6}.\frac{14}{8}.\frac{14}{6}.\frac{15}{9}\ldots\ldots\frac{21}{15}$$

<----------->

<-->

The only number of reds which gives rise to a whole number to one chance of a four ball break is four* as illustrated by the short dotted line. This multiplies out to **15**.

The long dotted line illustrates the clearance and multiplies out to **352800.**

(The clearance – in reverse – is....

black, pink, blue, brown, green, yellow, any colour, red, any colour, red, any colour, red, any colour, red)

For the sake of completeness the sequence is shown in full...

$$\ldots\ldots\quad\frac{14}{8}.\frac{14}{6}.\frac{15}{9}.\frac{15}{6}.\frac{16}{10}.\frac{16}{6}.\frac{17}{11}.\frac{17}{6}.\frac{18}{12}.\frac{18}{6}.\frac{19}{13}.\frac{19}{6}.\frac{20}{14}.\frac{20}{6}.\frac{21}{15}$$

*Only for n = 4 does a whole number result from $\dfrac{4+n}{6}.\dfrac{5+n}{n-1}.\dfrac{5+n}{6}.\dfrac{6+n}{n}$

52 SQUARES AND CUBES

Answer: 6, 10, 384 and 640

Let one pair of numbers be x and y, where x>y. Then $x^2 - y^2 = a^3$ and $x^3 - y^3 = b^2$.

The first thing to note is that there are infinitely many such pairs. Given one pair, you can multiply each number of the pair by a 6th power (n^6), and it still works:

$(n^6x)^2 - (n^6y)^2 = n^{12} (x^2 - y^2) = n^{12}a^3 = (n^4a)^3$ and $(n^6x)^3 - (n^6y)^3 = n^{18} (x^3 - y^3) = n^{18}b^2 = (n^9b)^2$

The smallest power of 6 is $2^6 = 64$, so the search for a solution should start with x=10, 11, ..., 15 and y = 2, 3, ..., 9. If this yields a result, then multiplying by 64 will give a solution with two three-digit numbers (the next 6th power is $3^6 = 729$, and this will not give three-digit solutions). If it fails, we will need to try another method.

Fortunately, starting with x=10, we quickly find that x = 10 and y = 6 gives a solution:

$10^2 - 6^2 = 64 = 4^3$ and $10^3 - 6^3 = 784 = 28^2$

We can multiply each number by 64 to find another solution, namely 640 and 384:

$640^2 - 384^2 = 64^3$ and $640^3 - 384^3 = 14,336^2$

There are no other possible solutions (a computer check verifies this), so the four required numbers are 10, 6, 640 and 384.

53   SUPER STREET

Answer: 23 125 137 144 and 196

Firstly listing the super prime numbers:

11 23 29 41 43 47 61 67 83 89 101 113 131 137 139 151 157 173 179 191 193 197 199 etc.

listing super squares

36 81 100 121 144 169 196 etc.

and super cubes

125 512 1000 etc.

In the above lists there are 23 super primes, 7 super squares and 3 super cubes; therefore to restrict ourselves to 31 houses, we eliminate the two big cubes, leaving 199 as the house occupied by George and Martha. Ordering the houses numerically, we get

11 23 29 36 41 43 47 61 67 81 83 89 100 101 113 121 125 131 137 139 144 151 157 169 173 179 191 193 196 197 199

The key is to note that none of the above list ends in 2 or 8 and thus the arithmetical progression of the least significant digits must be 3 4 5 6 7 and their total must end in 5. If that is to be a perfect square, the previous digit must be 2. Thus the total must be 225 or 625. The only number ending in 4 is 144 and in 5 is 125. That gives 269 at least so 225 is ruled out as total leaving 625 as a certainty.

There are six 3's, 23 43 83 113 173 and 193 two 6's, 36 and 196 and 5 7's, 47 67 137 157 and 197.

Starting with 36, we add to 269 to give 305 and need 320 for the required total – nothing doing.

Try 196, we add to 269 to give 465 leaving 160 for the required total: 23 and 137; also 113 and 47 qualifies.

Thus we have two possibilities 23 125 137 144 and 196 or 47 113 125 144 and 196
Ordinal positions   2 17 19 20   29   7 15 17 20   29

So the first one is applicable and they live at 23 125 137 144 and 196.

Answer: 113 cm

3D view (L front, X back) of crystal piece. Edge lengths different whole numbers.
Part-faces abc, ABC, X(B-b)H are Pythagorean triangles – legs a, A and X <10 cm.

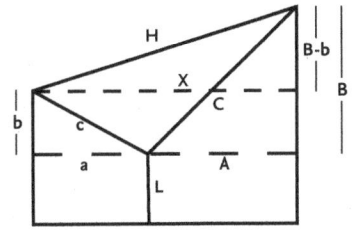

Only Pythagorean triples [P, Q, R] including single-figure values are
[3,4,5]; [6,8,10]; [9,12,15]; [5,12,13]; [7,24,25]; [8,15,17]; and [9,40,41]
by considering $P^2=R^2-Q^2=(R+Q)(R-Q)$ for P from 1 to 9
For example P=9, $P^2=81=81x1=27x3=9x9$
R+Q=81; R-Q=1 gives R=41, Q=40
R+Q=27; R-Q=3 gives R=15, Q=12
R+Q=9; R-Q=9 gives R=9, Q=0 invalid

So a, b, A, B, X and (B-b) are from P, Q options above; and c, C and H from R options above.

B not 3 or 4 and B not 40 (no valid 40-b values from above, so [9,40,41] eliminated)

So largest B=24 and so B-b not= 24 and 24-b is never 15, also B-b can't be 5, 7 or 9 else X>9. Thus B-b can only be 3, 4, 6, 8 or 12 so far. List **B-b** values valid so far:

6-3=3   8-4=4   12-8=4   12-6=6   12-4=8   15-12=3   15-3=12   24-12=12

Find abc, ABC, X(B-b)H sets for valid so far options

[B=6, b=3, A=8, a=X=4]; [B=8, b=4, A=6, a=X=3] – invalid – also A=a+X (no base triangle)

[B=12, b=8, A=5 or 9, a=6, X=3, H=5] – invalid – A=H or A=a+X (no base triangle)

[B=12, b=6, A=5 or 9, a=X=8] – invalid

[B=12, b=4, A=5 or 9, a=3, c=5, X=6] – invalid A=c or A=a+X (no base triangle)

[**B=15, b=12, A=8, C=17, X=4, H=5**, a=5] – invalid H=a (**or a=9, c=15 – valid so far**)

[**B=15, b=3, a=4, c=5, A=8, C=17**, X=5] – invalid X=c (**or X=9, H=15 – valid so far**)

For two **bold** options applied to the pentahedral form above their total perimeters are
P=a+A+X+c+C+H+(b+L)+(B+L)+L=(85+3L) and (76+3L). For P=odd prime; L is even for the former with [a=9, A=8, X=4, c=15 and b+L=12+even] a majority non-prime; L is odd for the latter with [a=4, A=8, X=9, H=15, b+L=3+odd] a majority non-prime – so both invalid

B=24, b=12, A=7, C=25, a=5 or 9 with c=13 or 15 and X=9 or 5 with H=15 or 13

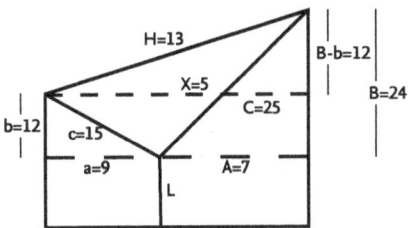

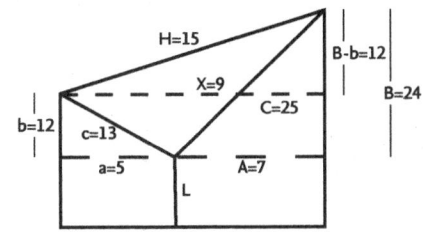

Perimeters of these valid variants are equal with a+A+X=21, c+C+H=53, b+B+3L=36+3L

Total perimeter P=110+3L=odd prime, Axial length=B+L=24+L<50, so L is odd<26

L not 1,3,5,7,9,13,15 or 25 (repeat edge lengths); L not 11 or 17 (P=143=11x13 or 161=7x23)

L not 21 (b+L=12+21=33, B+L=24+21=45) – minority prime

L=19, P=167 prime; and b+L=12+19=31, B+L=24+19=43 – majority prime

L=23, P=179 prime; and b+L=12+23=35=5x7, **B+L=24+23=47 – majority prime**

For other pentahedron, Pother=110+3N, with N<50-(B+L) – height < 50 cm

For L=19, B+L=43, N odd<7=1, 3 or 5, Pother=113 prime (or 119=7x17 or 125 – invalid)

For L=23, B+L=47, N odd<3=1, Pother=113 prime

55 BAKER'S WEIGHTS

Answer: 2, 3, 5, 9, 17 and 32 grams

If 1, 2, 3, …, n grams are weighed out the total weight is $S_n = n(n + 1)/2$ grams. Now $S_{44} = 44(44 + 1)/2 = 990$, so the apprentice expected to complete 44 weighings and be left with 10g in the bag, instead of which his final weighing exactly emptied the bag. In order to weigh out consecutive weights from 1-63 grams, a minimum of six baking weights are needed (1g, 2g, 4g, 8g, 16g and 32g). Consecutive weights from 1-44 grams can be weighed by replacing the 32g weight with a smaller one (eg. 1, 2, 4, 8, 16, 31) but not by replacing it with a larger one so the values engraved are 1g, 2g, 4g, 8g, 16g and 32g.

In the table below, the six weight columns contain a '0' if the weight is excluded and a '1' if it is included, so these columns show the expected weight in binary notation.

32g	16g	8g	4g	2g	1g	Expected	Total
0	0	0	0	0	0	0	0 (S_0)
0	0	0	0	0	1	1	1 (S_1)
0	0	0	0	1	0	2	3 (S_2)
0	0	0	0	1	1	3	6 (S_3)
0	0	0	1	0	0	4	10 (S_4)
1	0	1	0	0	1	41	861 (S_{41})
1	0	1	0	1	0	42	903 (S_{42})
1	0	1	0	1	1	43	946 (S_{43})
1	0	1	1	0	0	44	990 (S_{44})

The 1gm weight is alternately excluded and included so the 990_{10} total includes it 22 times. The 2g weight is alternately excluded twice and included twice so 990_{10} includes it 22 times. Repeating this process for the 4g, 8g, 16g and 32g weights enables the total weight to be written as follows:-

$S_{44} = 13 * 32g + 16 * 16g + 21 * 8g + 21 * 4g + 22 * 2g + 22 * 1g = 990$ grams

Adding 1g to any of the weights adds at least 13g to the total, so it is not possible to weigh out exactly 1,000g.

S_{43} includes one 32g, one 8g and one 4g weight less than S_{44} so it can be written:-

$S_{43} = 12 * 32g + 16 * 16g + 20 * 8g + 20 * 4g + 22 * 2g + 22 * 1g = 946$ grams

There are four ways of adding 1gm to some of the weights (shaded) to obtain a total of 1,000 grams:-

$12 * 33g + 16 * 16g + 20 * 9g + 20 * 4g + 22 * 3g + 22 * 1g = 1,000$ grams

$12 * 33g + 16 * 16g + 20 * 9g + 20 * 4g + 22 * 2g + 22 * 2g = 1,000$ grams

$12 * 33g + 16 * 16g + 20 * 8g + 20 * 5g + 22 * 3g + 22 * 1g = 1,000$ grams

$12 * 33g + 16 * 16g + 20 * 8g + 20 * 5g + 22 * 2g + 22 * 2g = 1,000$ grams

S_{42} includes one 32g, one 8g, one 2g and one 1g weight less than S_{43} so it can be written:-

$S_{42} = 11 * 32g + 16 * 16g + 19 * 8g + 20 * 4g + 21 * 2g + 21 * 1g = 903$ grams

There is only one way of adding 1gm to several weights to obtain a total of 1,000 grams:-

$11 * 32g + 16 * 17g + 19 * 9g + 20 * 5g + 21 * 3g + 21 * 2g = 1,000$ grams

S_{41} includes one 32g, one 8g, one 2g and one 1g weight less than S_{42} so it can be written:-

$S_{41} = 10 * 32g + 16 * 16g + 18 * 8g + 20 * 4g + 20 * 2g + 21 * 1g = 861$ grams

Adding 1g to every weight gives a total of 930 grams, which is insufficient, and S_1 to S_{40} are smaller than S_{41} so it is not possible to obtain a total of 1,000g for any other weighing.

The six weights can only be deduced if told that five weights are 1g heavier than their engraved weight, hence the baking weights are 2, 3, 5, 9, 17 and 32 grams.

56 CONSECUTIVE SUMS

Answer: 289

For any number N, we want to know how many ways there are of adding consecutive whole numbers to equal N (I'll call them constructions). We need to consider cases where the number of summands (numbers to be added) is (1) odd or (2) even and combine them.

1) The average of the summands must be a whole number. If the average is a, then
   $N = (a-n)+...+(a-1)+a+(a+1)+...+(a+n) = a(2n+1)$, where the number of summands is 2n+1
   Clearly a-n must be > 0 for all the numbers to be positive, so N>n(2n+1)
   Now let x=2n+1 be any odd factor of N, so N>x(x-1)/2
   Then for any odd factor x of N, there will be a corresponding construction with an odd number of summands, provided that x(x-1)<2N.

2) The average of the summands must be a whole number + 0.5. If the average is a, then
   $N = (a-n+0.5)...+(a-0.5)+(a+0.5)+...+(a+n-0.5) = 2an$, where the number of summands is 2n
   Clearly, a-n+0.5>0 so n<a+0.5 and N<a(2a+1)
   Now a must be a whole number + 0.5, so 2a must be an odd number, and N has an odd factor.
   Let x=2a be any odd factor of N, so N<x(x+1)/2.
   Then for any odd factor x of N, there will be a corresponding construction with an even number of summands, provided that 2N<x(x+1)

Combination

Every odd factor x of N gives rise to a construction, provided that either x(x-1)<2N or 2N<x(x+1). Now let N=xy where x is odd and y is a whole number. The inequalities are then x-1<2y and 2y<x+1. The only value for which these could both be true is 2y=x, but this is impossible because x is odd. Also, 2y <= x-1 and 2y >= x+1 cannot both be true. Therefore, precisely one of the above inequalities is always satisfied, and we have found all the possible constructions.

Therefore, the number of constructions is equal to the number of odd factors of N ignoring the factor 1 (as there will only be one summand). If N is prime, there is just one odd factor and if N has two or more different prime factors (e.g. p and q), then it will have at least three odd factors (p, q and pq). If N is a square of a prime number (e.g. p), then it has just two odd factors (p and p^2), while for cubes etc there will be three or more odd factors. We only are interested in odd factors, so multiplying N by 2 any number of times doesn't change the number of constructions. Therefore, for Amelia and Ben (having just two constructions each), N must be of the form $p^2 2^n$. The table on the next page shows all the possible 3-digit values.

Prime (p)	p^2	$p^2 \times 2$	$p^2 \times 4$	$p^2 \times 8$	$p^2 \times 16$	$p^2 \times 32$	$p^2 \times 64$
3	9	18	36	72	144	**288**	576
5	25	50	100	200	400	800	
7	49	98	196	392	784		
11	121	242	484	968			
13	169	338	676				
17	**289**	578					
19	361	722					
23	529						
29	841						
31	961						

Examination of the table reveals that there is just one occurrence of consecutive 3-digit numbers, 288 and 289. So Ben's number is 289.

Answer: 1854

The answer involves a permutation of elements in a set such that no element appears in its original, or correct, position. Such a permutation is called a derangement.

Let the number of derangements for a set of n items be d(n). For small values of n, d(n) can easily be shown as being –

			With items in their
For n = 1 d(1) = 0	-	1	correct or original positions shown in bold and boxed when all items are otherwise.
For n = 2 d(2) = 1	-	**12**, 21	
For n = 3 d(3) = 2	-	**123**, 132, 213, 231, 312, 321	

With items in their correct or original positions shown in bold and boxed when <u>all</u> items are otherwise.

For higher values of n, d(n) is more laborious to find but can be calculated as explained below.

Suppose that there are *n* people who are numbered 1, 2, ..., n. Let there be n drinks also numbered 1, 2, ..., n. We have to find the number of ways in which no one gets the drink having the same number as their number. Let us assume that person 1 receives drink *i* where *i* is not 1 i.e. the wrong drink. There are n – 1 ways for this first person to receive this wrong drink. There are now two possibilities, depending on whether or not person *i* is given drink 1:

1) Person *i* does not receive drink 1. This case is equivalent to solving the problem with n–1 persons and n–1 drinks: each of the remaining n–1 drinkers have precisely 1 drink they can't receive from among the remaining n–1 drinks. This yields the first term of the sum: (n-1)*d(n-1).

2) Person *i* has drink 1. Now two drinkers/drinks have been accounted for leaving n–2 drinkers and n–2 drinks. This yields the second term of the sum: (n-1)*d(n-2).

Putting the two terms together gives the following equation.

   d(n) = (n-1){d(n-1)+d(n-2)}

Substituting d(1) = 0 and d(2)=1 gives d(3) = 2, repeating the substitution for d(2) and d(3) gives d(4) =9. Continuing this repetition gives the following results-

n	d(n)
1	0
2	1
3	2
4	9
5	44
6	265
7	1,854
8	14,833
9	133,496

d(7) = 206*d(4), which gives the answer of 1854.

[d(n) cannot be 206*d(n-2); the nearest possibilities are d(14) = 182*d(12) - 13 and d(15) = 210*d(13) + 14, and for d(n) to be 206*d(n-1), n would have to be more than 20.]

Answer: 9

The pieces will be some or all of:

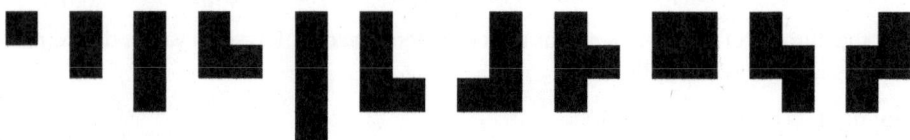

With a maximum area of 37cm² and more than one square of size at least 2x2, the biggest possible square is 5x5.

I have made a set of squares with total area at most 37cm². So the total area of my squares will be

4+9 or 4+16 or 4+25 or 9+16 or 9+25 or 4+9+16.

My set of pieces can make more than one different sets of squares, so the total area of my pieces will occur more than once in that list. Therefore it must be

29 = 4+25 = 4+9+16

How can we make a 2x2 and a 5x5 square? The 2x2 square must use two pieces of areas 1 and 3. So the pieces in the 5x5 square must be chosen from those of areas

2   3   4   4   4   4   4   4   4.

This is only possible with seven pieces of areas 2, 3, 4, 4, 4, 4 and 4. So overall my set must consist of nine pieces.

We illustrate two possible sets of squares using the same nine pieces:

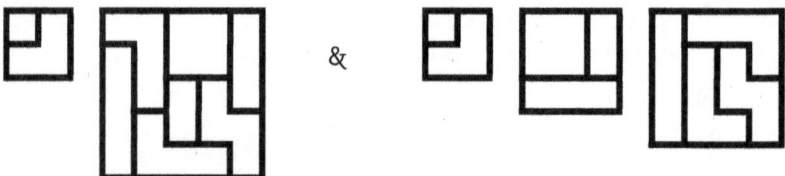

&

Answer: 6, 15, 20, 34

Given that "excluding 1 and the numbers themselves", each value shares one factor with another one, no value can be prime.

One odd value and shared factor criteria mean lines have values in either of two formats:-

[V1=even, V2=even, V3=odd, V4=even] or [V1=even, V2=odd, V3=even, V4=even]

Tabulate valid options: * means invalid

Note, V2=10 [leads to V3=25] or 16 [shares factor 2 with V1 and V3]; and V3=25 [shares factor 5 with V2 and V4] or 27 [shares factor 3 with V2 and V4] – invalid

V1 (even)	4	4	4	6	6	6	6	8	8	8
[share factor]	2	2	2	2/3*	2	3	2/3*	2/4*	2	2
V2	12	14	18	12	14	15	18	12	14	18
[share factor]	3	7	3		7	5			7	3
V3	21	21	21		21	20			21	21
[share factor]	7	3	7		3	2			3	7
V4 (even)	35*	30 or 36	35*		30 or 36	34 or 38			30 or 36	35*
[share factor]		2			2/3*	2			2 or 2/4*	
V1 (even)		4			6	6			8	

Valid lines:- [4, 14, 21, 30]; [4, 14, 21, 36]; [6, 15, 20, 34]; [6, 15, 20, 38]; [8, 14, 21, 30]

Just two of these have the same sum, 75, so one line is 4, 14, 21, 36 and other line is 6, 15, 20, 34. The top line has highest number (36).

So the bottom line is 6, 15, 20, 34

Answer: 20

From a., if there were n teams in Div I then the total number of matches is

> Div I $n(n-1)$ Div II $n(n+1)$

Each drawn match reduces the total points by one. There must be fewer draws in Div I for there to be an equal no. of points awarded.

From d., $D = N d$

.where D is the no. of draws in Div II

> d I
> N is an integer.

From c.,

> In Div I there were d draws and D wins.

Points awarded in Div I $= 2d + 3D$
(draws) (wins)
$= 2d + 3n(n-1) - 3d$

Points awarded in Div II $= 3[n(n+1) - D] + 2[n(n-1) - d]$
(wins)   (draws)

Equating and applying $D = n(n-1) - d$,

$$3n(n+1) = 3D + 2[n(n-1) - D]$$

$$D = \frac{n(n+5)}{2} \quad \text{and} \quad d = \frac{n(n-7)}{2}$$

Using $D = N d$

$$n(n+5) = N n(n-7)$$

$$n[n(N-1) - (7N+5)] = 0$$

$$n = \frac{7N+5}{N-1}$$

Only for N = 2 does d become a 3 fig. number so n = 19

(N	n	d
2	19	114
3	13	39)

Division	Teams	Draws	Wins	Points
I	19	114	228	912
II	20	228	152	912

Division II contains 20 teams.

61   TRIANGULAR CARD TOWER

Answer: 355

It can quickly be seen that for one level 2 cards are needed, for two levels 7 cards (four support cards on the first level, one horizontal card and then two leaning cards on the second level), for three levels 15 cards, four 26 cards etc.

This is a second order arithmetical progression with the following formula for the number of cards for the nth stage –

$$= 2 + 5(n-1)!/(n-2)! + 3(n-1)!/2(n-3)! = 2 + 5(n-1) + 3(n-1)(n-2)/2 = n(3n+1)/2 \qquad (1)$$

Let N be the number of bottom levels of larger cards and n be the total number of levels (big and small cards).

The total number of large cards used is therefore from (1) $n(3n + 1)/2 – (n – N)\{3(n – N) + 1\}/2$
Which reduces to – Number of Large Cards Used = $N(6n – 3N + 1)/2$ $\qquad (2)$
Let the length of the bigger cards be l
The vertical height of the cards is, by Pythagoras, equal to $\sqrt{\{l^2 - (0.56\ l\ /2)^2\}} = 0.96l$
The height of all the cards, larger and smaller is $0.96lN + 0.96l *0.75*(n - N)$
$= (0.24N + 0.72n)l$ $\qquad (3)$
As per the teaser $(0.24N + 0.72n)l = 1428$ which reduces to $3n = 5950/l – N$ $\qquad (4)$
N is also the number of 53 card packs equaling the number of large cards in the bottom layers.
So from (2) above $N(6n – 3N + 1)/2 = 53N$
Substituting for 3n from (4) and dividing by N gives $2 * 5950/l – 3N + 1 = 2 * 53N$
Which reduces to $l(21 + N) = 2380$
Using factors of 2380 greater than 21 gives the following values of N and l

Factor	N	l
28	7	85
34	13	70
35	14	68

As the bigger cards are longer than 70 their length must be 85, and there are 7 bottom levels.
Substituting $l = 85$ and N = 7 in equation (3) gives the height of the towers (h) as being
$h = 85(0.24 * 7 + 0.72n) = (306n + 714)/5$
from this it can be seen that there are integer solutions for h when n = 1 and then for every value of n increased by 5. The following table shows these results –

h	n
204	1
510	6
816	11
1,122	16
1,428	21
1,734	26

The number of card levels therefore has to increase from 21 to 26 for the height of 1,428 to reach the next integer height of 1,734.

From equation (1) the total number of cards has to increase from 672 to 1,027 an increase of 355 cards.

Answer: 12X523

```
1  2  3  4  5  6
   T  H  R  E  E
+  T  H  O  U  S
+  A  N  D  T  H
T  E  A  S  E  R
```

From col 1, T = 1.

From col 2, since 0 is not used in the sum E = 2, with A = 9 (with a carry of 2 from col 3) or A = X (with a carry of 1 from col 3). If A = 9 the maximum possible value of (H+H+N) in col 3 is 26 (X+X+8); to give A = 9 this would require a carry of 3 from col 4, which is impossible. So A = X.

If there is no carry from col 6 U = X (col 5); but A = X. Since E = 2 the carry from col 6 cannot be 2, so must be 1, with U = 9, and the carry from col 5 to col 4 is 1. Since T = 1, E = 2 and 0 is not used there must be a carry from col 4 to col 3.

The following values of H and N would be possible for col 3 if there is a carry of 2 from col 4: H = 8, N = 3 (largest remaining: 7, 6, 5); H = 7, N = 5 (largest remaining: 8, 6, 4); H = 6, N = 7 (largest remaining: 8, 5, 4). But in each case even if R, O and D are the "largest remaining" their sum (plus the carry of 1 from col 5) is not great enough to give a carry of 2 from col 4.

So the carry from col 4 to col 3 is 1, with these possible values: H = 8, N = 4; H = 7, N = 6; H = 6, N = 8.

If H = 8 and N = 4 the digits left are 3, 5, 6, 7.

From col 6 S could be 6 with R = 5 (and O & D 3 & 7), or S could be 7 with R = 6 (and O & D 3 & 5), but in col 4 5+3+7+1 (carry) = 15, not 16; 6+3+5+1 (carry) = 14, not 17.

If H = 6 and N = 8 the digits left are 3, 4, 5, 7.

From col 6 S could be 7 with R = 4 (and O & D 3 & 5), but in col 4 4+3+5+1 (carry) = 12, not 17.

If H = 7 and N = 6 the digits left are 3, 4, 5, 8.

From col 6 S could be 5 with R = 3 (and O & D 4 & 8), and in col 4 3+4+8+1 (carry) = 15 as required.

O and D are interchangeable in the following solution, but this does not matter so long as solvers are not required to identify these two letters in their answer.

```
   1  7  3  2  2
   1  7  4  9  5
   X  6  8  1  7
1  2  X  5  2  3
```

Answer: 3 in.

'Square symmetry' pattern from 13 tiles, must use a central square, and 4 each kites, rectangles, rhombuses (12 tiles) arranged symmetrically with their reflection symmetry axes along a perpendicular pair of symmetry axes of the square, example as illustrated below.

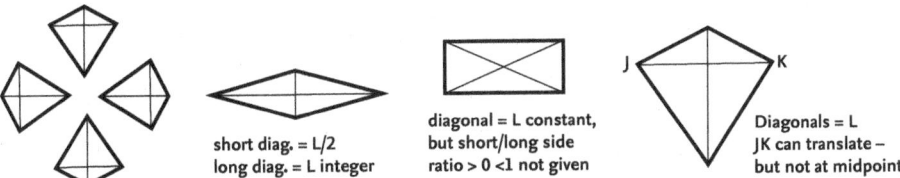

short diag. = L/2
long diag. = L integer

diagonal = L constant,
but short/long side
ratio > 0 <1 not given

Diagonals = L
JK can translate –
but not at midpoint

With fixed diagonal length L, rectangle longer side length must be <L

Check only 'vertex touching' schemes below in which the rhombuses must contact each other type, including the single centrally placed square. A rectangle vertex to square vertex can't give overall square symmetry for a pattern as specified as may easily be seen by sketching.

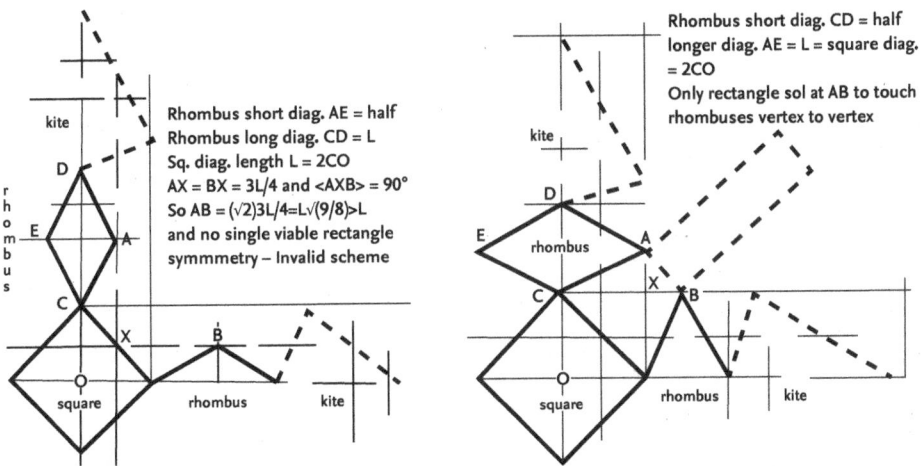

Rhombus short diag. AE = half
Rhombus long diag. CD = L
Sq. diag. length L = 2CO
AX = BX = 3L/4 and <AXB> = 90°
So AB = (√2)3L/4=L√(9/8)>L
and no single viable rectangle
symmmetry – Invalid scheme

Rhombus short diag. CD = half
longer diag. AE = L = square diag.
= 2CO
Only rectangle sol at AB to touch
rhombuses vertex to vertex

In the right hand diagram AX=BX=L/4 and <AXB>=90 °, so AB=(√2)L/4 (integer L>=1)
AB is rectangle short side. Rectangle diagonal=L. So by Pythag. Theorem we get rectangle long side RLS=√(L^2-AB^2)=L√(1-1/8)=L√(7/8)

Tabulate rectangle sides to 2 dp

L=	1	2	3	4	>4
AB=	0.35	0.71	**1.06**	1.41 (too far over 1)	>1.41
RLS=	0.94	1.87	**2.81**	>2.81	

Only rectangle option with a side slightly over 1 in. is for L=3 in.

64 SHORT-CUT

Answer: 32 cm

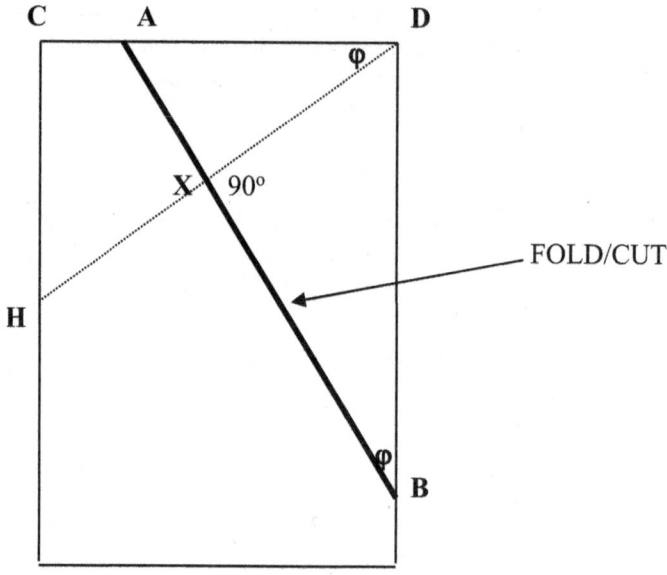

Let t = tanφ (<1 to ensure that A is in CD in order to get a triangle after the cut), so cosφ = $1/\sqrt{(1 + t^2)}$ and sinφ = $t/\sqrt{(1 + t^2)}$.

Then

$$CH = 24t \quad HX = XD = 12\sqrt{(1 + t^2)},$$
$$AX = 12t\sqrt{(1 + t^2)}, \quad XB = (12\sqrt{(1 + t^2)})/t$$

Area of triangle ADB = one third of the rectangle = 48t.24/3 = 384t (*)

But the area of ADB is ½ XD(AX + XB) which is

$$6\sqrt{(1 + t^2)}[12t\sqrt{(1 + t^2)} + (12\sqrt{(1 + t^2)})/t] = 72[t(1 + t^2) + (1 + t^2)/t] = 72(1 + t^2)^2/t \ (*)$$

So equating these two (*) and tidying up gives
$$16t^2 = 3(1 + t^2)^2 \quad \text{or} \quad 3 - 10t^2 + 3t^4 = 0$$

Therefore $t^2 = 1/3$ (3 being ruled out) and tanφ = t = $1/\sqrt{3}$.

So CH = $8\sqrt{3}$, HX = XD = $8\sqrt{3}$, AX = 8, XB = 24 and the length of the cut is **32**.

[In fact this makes DA = 16 and DB = $16\sqrt{3}$, making B the corner of the rectangle.]

Answer: 1903, 1936, 1969

The number of factors of an integer is usually an even number because they are naturally paired.

For example 12 has $(1, 12), (2, 6), (3, 4) = 6$ factors. The exception is a perfect square. For example 16 has $(1, 16)$, $(2, 8)$ and 4. The sum of the number of factors of the dates is odd, so one of them must be a perfect square.

Since $43^2 = 1849$ and $45^2 = 2025$ the only available square is $1936 = 44^2 = 2^4.11^2$.

Since the h.c.f. of each pair of dates is an odd prime that prime must be 11, and the dates are all divisible by 11 and drawn from the set 1903, 1914, 1925, 1936, 1947, 1958, 1969, 1980, 1991.

Since the highest common factor of any pair of dates is an odd prime (that is now 11), the common difference in the equally spaced dates is either 11 or 33; 22 would create a common factor of non-prime 22 between 1936 and another date, and 44, 55 ... are too big.

Since the number of coins is an odd prime there are either 3, 5 or 7 dates. If there are 5 or 7 dates the common difference must be 11, but 1914, 1936, 1958 and 1980 are all divisible by 2 and any 5-date or 7-date sequence must contain two of those dates, giving a common factor between the pair of non-prime 22.

So there are three dates, with a common difference of 11 or 33, and containing 1936. The possibilities are:
1914, 1925, 1936: but this has a pair of even dates with a common factor of non-prime 22.
1925, 1936, 1947
1936, 1947, 1958: but this has a pair of even dates with a common factor of non-prime 22.
1903, 1936, 1969

Now we have to find the sum of the number of factors in the dates by prime factorising them. This is easy since each date has a factor of 11. The number of factors is the product of the indices of the prime factorisation each increased by 1, since if the index is n that prime might appear 0, 1, 2. ... n times in a factor's prime factorisation. The number of factors calculation is given in the brackets.

$1925 = 11.5^2.7$ $(2.3.2 = 12)$ $1936 = 2^4.11^2$ $(5.3 = 15)$ $1947 = 11.3.59$ $(2.2.2 = 8)$
(sum $12 + 15 + 8 = 35$)

$1903 = 11.173$ $(2.2 = 4)$ $1936 = 2^4.11^2$ $(5.3 = 15)$ $1969 = 11.179$ $(2.2 = 4)$
(sum $4 + 15 + 4 = 23$)

Since 35 is not an odd prime as required, and 23 is an odd prime, the dates can only be 1903, 1936, 1969.

The reason for the title 'All that glitters...' is that a "1936 sovereign" is famously counterfeit because none were struck in the year of the abdication of Edward VIII in December 1936. A handful of paradoxical 1937 Edward VIII sovereigns were struck for technical reasons and are among the most valuable coins in existence, one selling for £516,000. No sovereigns were struck from 1969-1973, during decimalisation, so that coin is counterfeit too.

66 GOING UP

Answer: 6 and 7

There are two ways in which the remaining fixtures can help a team guarantee promotion with two wins:

1) The team plays against two of its rivals. In this way, two possible rivals are eliminated, i.e. both of them will finish lower in the league because they can't win both of their games.

2) The team's rivals play against each other, taking points off each other. If there are two such teams, only one can win both its remaining games. If there are three such teams playing against each other (e.g. AvB, AvC, BvC), then again only one of them can win both its games. For more than three teams playing against each other (e.g. AvB, AvC, BvD, CvD), then it is possible for two teams (e.g. A and D) to win both their games.

In order for a team to guarantee promotion by winning its two remaining games, there can be no more than one other team which could win both its games (and deprive the first team of promotion on goal difference). Therefore, there can be no more than three rivals playing against each other in (2), and no more than two rivals can be eliminated in (1). Therefore, the maximum number of teams with the same number of points that allows for a team to guarantee promotion with two wins is six.

We need to find a set of final fixtures that allows <u>just one team</u> to guarantee promotion from the six with two wins. First of all, we must have three of them playing against one another (AvB, AvC, BvC), as in (2). Next, we must have one of the other teams playing against two rivals (DvE, DvF), as in (1). If EvF is a remaining fixture, then any of the six teams could guarantee promotion with two wins, so we must have EvG and FvH (where G and H are any teams not in the six) and the full set of fixtures involving any of the six is: AvB, AvC, BvC, DvE, DvF, EvG, FvH

Then
A could be overtaken by E and F on goal difference
B could be overtaken by E and F on goal difference
C could be overtaken by E and F on goal difference
E could be overtaken by A and F on goal difference
F could be overtaken by A and E on goal difference
D could not be overtaken by two teams, because E and F are eliminated, and only one of A, B and C can win both its games, so D is the <u>one team</u> which can guarantee promotion with two wins.

There are six teams together on points with seven remaining matches involving any of them.

Answer: 37mm

Placed tube radii **r1=11** and r2, r3, r4, r5, r6, r7, r8, r9 permed from 13, 17, 19, 23. 29, 31, 37 and 41 (r2 not=41)

11+r2=multiple of one or more of r3 to r9. 7 to test – only 11+23=34=2x17 works, so **r2=23**

r1+r2+r3=34+r3 is a multiple of r1, r2 or r3. 7 to test – only **r3=17**, 51=3x17 works

Now 51+r4 is multiple of r1, r2, r3 or r4. With 6 to test, two work – r4=37 or r4=41 – giving 51+37=88=8x11 or 51+41=92=4x23

For r4=37, 88+r5 is multiple of r1, r2, r3, r4 or r5. Five to test, only r5=31, 119=7x17 works, but 119+r6 is multiple of r7, r8 or r9 from 13, 19, 29 or 41 – none work

So, **r4=41** and 92+r5 is multiple of r1, r2, r3, r4 or r5. From 5 values to test, two work – r5=29 or r5=31 – giving 92+29=121=11x11 or 92+31=123=3x41

For r5=29, 121+r6 is a multiple of an unused radius from 13, 19, 31 or 37. Only r6=31 works – giving 121+r6=121+31=152=8x19

For r5=31, 123+r6 is a multiple of an unused radius from 13, 19, 29 or 37. Only r6=29 works, giving 123+r6=123+29=152=8x19

With the same aggregate radius sum=152, r5=29, r6=31 or r5=31, r6=29 options work

Now 152+r7 is multiple of r1, r2, r3, r4, r5, r6 or r7. With 13, 19, 37 to test, two work – r7=13 or r7=19 – giving 152+13=165=15x11 or 152+19=171=9x19

For r7=13, 165+r8 is multiple of r1, r2, r3, r4, r5, r6, r7 or r8. With 19 and 37 to test, only r8=19 works, with 165+19=184=8x23, leaving **r9=37**; (165+37=202=2x101 invalid)

For r7=19, 171+r8 is multiple of r1, r2, r3, r4, r5, r6, r7 or r8. With 13 and 37 to test, only r8=13 works, with 171+13=184=8x23, leaving **r9=37**; (171+37=208=16x13 invalid)

Notwithstanding the possible transpositions of r5, r6 and r7, r8 - **r9=37mm**

Below is the minimal gap arrangement. 29, 31 and 13, 19 swaps still give <1mm gap

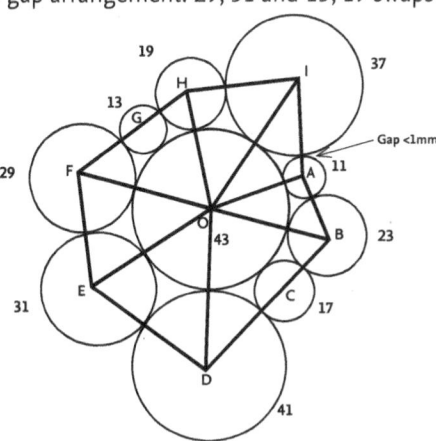

Purists can confirm this using cosine rule iterations on triangles in nonagon ABCDEFGHI

Answer: 2, 3, 41, 58, 69, 70

The full list of possible numbers is

	10		30				70
		21	31	41	51	61	71
2				42		62	
3	13	23		43	53		73
	14		34				74
5	15		35			65	
6		26		46			
7	17		37	47	57	67	
			38		58		78
	19	29	39		59	69	79

The sum of the digits is 45.
The sum of the numbers must be divisible by 9.
The digits 9, 8 and 0 must be the units digits for three of the four two-digit numbers.
The only possible perfect powers are 12^2 (144), 15^2 (225), 6^3 (216) and 3^5 (243).

	144	225	216	243
Tens	11	20	19	22
Units	34	25	26	23
	6 7 10 29 34 58	2 5 30 41 69 78	2 6 30 41 58 79	2 3 41 58 69 70
	6 7 14 29 30 58	2 5 38 41 69 70	2 6 30 41 59 78	
			2 6 38 41 59 70	
			2 6 39 41 58 70	
			2 3 10 58 69 74	
			2 3 14 58 69 70	

Given that you can work out the numbers from knowing the sum, then the sum must be
243 and the numbers are   **2   3   41   58   69   70.**

69 PAVING STONES

Answer: 5520

At the centre there is a single row of square yellow blocks, let this be x blocks long. Working out from the centre, and numbering the bands from the centre, the number of blocks in each band is –

Band 1	(yellow)	x	
Band 2	(grey)	2x + 6	)
Band 3	(grey)	2x + 14	) Arithmetic Progression (AP) with
Band 4	(grey)	2x + 22	) difference of 8.
Band 5	(red)	2x + 30 etc	)

For all bands, except band 1, the number of blocks in each band is $2x + (8n - 10)$ (1)
where 'n' is the band number.

Let 'R' be the number of times the pattern yellow/grey/red repeats before filling the rectangle. The number of yellow, grey and red blocks in each pattern repeat are also APs (except for the first innermost yellow line of blocks). There are 8 block bands in each pattern repeat.

For yellow, for which the pattern comprises of a single band, starting from the centre the number of blocks in successive patterns is from (1) band 1 = x, band 9 = 2x+62, band 17 = 2x + 126 etc

Above band 1 this is an AP with the first term 2x + 62 and difference 64 (the AP difference for each band irrespective of colour is 8 and the number of bands in each pattern is 8 so 8x8=64).

Using the sum of an AP formula of Sum = $(n/2)\{2a + (n - 1)d\}$ where 'n' is the   (2)
number of terms, 'a' is the first term and 'd' is the difference gives –

Sy (Sum of Yellow Blocks) = $x + \{(R - 1)/2\}\{2(2x + 62) + (R - 2)64\}$
(the first innermost band of yellow blocks is outside the AP sequence and so has to be separately added and the number of terms reduced by 1). This reduces to
Sy = $32R^2 + 2xR - x - 34R + 2$

For the red blocks the band numbers are: first pattern (R=1) band 5 = 2x + 30, band 6 = 2x + 38, band 7 = 2x + 46, band 8 = 2x + 54; second pattern (R=2) band 13 = 2x + 94, etc
The initial term of the red block AP (R=1) is (2x+30) + (2x+38) + (2x+46) + (2x+54) = 8x + 168
The difference between the Red block AP is 256 (8 (difference between each band) x 8 (number of bands in each pattern) x 4(number of red block bands in each pattern).
From (2) this gives Sr (Sum of Red Blocks) = $(R/2)\{2(8x + 168) + 256(R - 1)\}$
This reduces to Sr = $R(128R + 8x + 40)$
But Sr = 5 * Sy so $R(128R + 8x + 40) = 5(32R^2 + 2xR - x - 34R + 2)$
Which reduces to $32R^2 + 2xR - 5x - 210R + 10 = 0$ (3)

The number of red blocks in the outermost band is from (1) $2x + (8n - 10)$ but as there are 8 bands in each pattern n = 8R so
$2x + (10n - 10) = 2x(8*8R - 10) = 402; x = 206 - 32R$ (4)
Substituting for x in (3) and simplifying gives $16R^2 - 181R + 510 = 0$
Using the Quadratic formula this gives the two solutions for R as 6 and 5.3125
As R has to be a whole number R = 6 and so n = 48 and from (4) x = 14

Using R=6 and x=14 gives Sr = 5520.

Answer: 5832

Three identical wheels – all fair
[Total zones around rim (Z)] X [Two-figure degrees angle per zone (A)] = 360°
So A is two-fig. factor of 360, Z=360/A, tabulate options

A	90	72	60	45	40	36	30	24	20	18	15	12	10
Z	4	5	6	8	9	10	12	15	18	20	24	30	36

All chances for 'three of same fruit' are of form 1/integer – for cherries this $=(1/Z)^3=1/(Z^3)$

For some other fruit with F repeats the chance is $(F/Z)^3=(F^3)/(Z^3)=1/integer$
So (F^3) is a factor of (Z^3) and F is a single-fig. factor of Z
By prime factorization if $F=[P^x][Q^y]...$ and $Z=[P^a][Q^b]...[...]$ then $F^3=[P^{3x}][Q^{3y}]...$
and $Z^3=[P^{3a}][Q^{3b}]...[3...]$
$(Z^3)/(F^3)=integer=[P^{3(a-x)}][Q^{3(b-y)}]...[3...]$ with a>=x, b>=y, etc.
So $Z/F=[P^{(a-x)}][Q^{(b-y)}]...[...]=integer$ and so F is factor of Z

Z=1+Sum of (all or a subset of) possible F values

Z(ones)	Pr. Fact.	1 and F options	Z=1+sum F vals	Diff. fruits incl. cherry	For example (excl. cherry)
4	2x2	1, 2	no		
5	5	1	no		
6	2x3	1, 2, 3	yes	3	fig, lime
8	2x2x2	1, 2, 4	no		
9	3x3	1, 3	no		
10	2x5	1, 2, 5	no		
12	2x2x3	1, 2, 3, [4], 6	yes [exclude]	4	fig, lime, date
15	3x5	1, 3, 5	no		
18	2x3x3	1, 2, [3], 6, 9	yes [exclude]	4	fig, lime, date
20	2x2x5	1, 2, 4, 5	no		
24	2x2x2x3	1, 2, 3, 4, 6, 8	yes	6	fig, lime, date, plum, kiwi
30	2x3x5	1, 2, 3, 5, 6	no		
36	2x2x3x3	1, 2, 3, 4, 6, 9	no		

Calculate chance fractions for various fruits for Z=6, 12, 18 and 24

Z	cherry	fig	lime	date	plum	kiwi
6	(1/6)^3	(2/6)^3	(3/6)^3			
	=1/even	=1/odd	=1/even			
12	(1/12)^3	(2/12)^3	(3/12)^3	(6/12)^3		
	=1/even	=1/even	=1/even	=1/even		
18	**(1/18)^3**	**(2/18)^3**	**(6/18)^3**	**(9/18)^3**		
	=1/even	**=1/odd**	**=1/odd**	**1/even**		
24	(1/24)^3	(2/24)^3	(3/24)^3	(4/24)^3	(6/24)^3	(8/24)^3
	=1/even	=1/even	=1/even	=1/even	=1/even	=1/odd

Only **Z=18** has more than one odd denominator in chance fractions, so chance for
3 cherries is 1/(18^3)=1/5832

Answer: 0, 2, 4, 4, 3, 1, 1

The possible totals of points (mine + grandson's) scored after each round are

7/17/27   13/23/.../53   18/28/.../78   22/32/.../102   25/35/.../125   27/37/.../147   28/38/.../168

On four or more occasions the cumulative totals are AB/BA which add to 11(A+B); i.e. 3 or more times 11. The only possible totals in the above list are 33, 55, 77 and 88. Therefore the total points after each round are

7/17/27	33	38/48	42/52	55	77	88
	scores 12/21			14/41 or 23/32	16/61, 25/52 or 34/43	17/71, 26/62 or 35/53

For the cumulative scores to be N/2 and N in some round (with a larger lead earlier) the total must be divisible by 3 and so it must be 42 or 48, with cumulative scores 14/28 or 16/32 respectively. But the 14/28 pair would have to be followed by 14/41, giving bigger leads than 14 on two later occasions (the 14/41 and the final totals). So we have

7/17/27	33	48	52	55	77	88
	scores 12/21	scores 16/32		scores 23/32	scores 34/43	scores 35/53

Then the total of 52 must be from scores 20/32.

With scores 16/32 my grandson's lead was 16. This lead is beaten once earlier and once later (35/53). The earlier occasion can only be after the first round with scores 0/17. So overall the cumulative totals are:

0/17   12/21   16/32   20/32   23/32   34/43   35/53

The full story is:

	round 1 (7 coins)	round 2 (6 coins)	round 3 (5 coins)	round 4 (4 coins)	round 5 (3 coins)	round 6 (2 coins)	round 7 (1 coin)
No. of heads	0	2	4	4	3	1	1
My score	0	12	4	4	3	11	1
His score	17	4	11	0	0	11	10

Answer: (a) 5 (b) 6 and 8

The "scoring area" for each hole is 4cm less than its width because the ball's width is 4cm. The width of the scoring area is 11-s, where s is the score for the hole. For example, for hole 8 (width 7cm), the centre of my ball has a 3cm window to pass through in order to score. The "dead area" where I don't score covers 6 cm on one side of the scoring area (2cm at the right-hand side of the hole, 2cm gap then 2cm at the left-hand side of the next hole), then there is another scoring area and so on.

The "aiming area" within which my ball's centre arrives covers 24cm, so for each scoring area that lies within the aiming area, I score on average $s*(11-s)/24$, so the best scoring holes are 5 and 6, followed by 4 and 7, then 3 and 8 etc.

The aiming area of 24cm can cover the scoring area of two holes and some of another. It won't cover the scoring area of more than three holes.

Assume the holes covered by my aiming area are A, B and C in that order from left to right. Also assume that C>A. I aim at the middle of B and for this to give a maximum average score, my aiming area must extend to the right-hand end of C's scoring area and to somewhere in A's scoring area (which is larger than C's since C>A). If I aim further left, the average score is lower because more of A's scoring area is included at the expense of C's. If I aim further right, the average score is lower because less of A's scoring area is included but no more of C's is included (as all of it is already included).

From the centre of B to the right-hand end of C's scoring area, you have half of B's scoring area, a dead area then all of C's scoring area, so the width of this is $(11-B)/2 + 6 + (11-C)$. This must equal 12 (half my aiming area). The equation reduces to $B + 2C = 21$. The only possibility is B=5 and C=8. B's scoring area is 6cm, while C's is 3cm. By symmetry, my aiming area will also include 3cm of A's scoring area. The average score will then be $(5*6+8*3+A*3)/24$, and A=6 gives an average score of 3.

I therefore aim at hole 5, with 6 and 8 either side.

[The order of the holes could be 7 2 6 5 8 4 1 3. Aiming anywhere other than the middle of 5 would give a lower score: we have used the two best-scoring holes 5 and 6, and 2 is lower-scoring than 8, so aiming near the middle of 6 will give a lower score. For completeness, the Appendix shows the average scores for every possible aiming point.]

[Appendix]

This shows the aiming point distance in cm from left-hand edge of hole 7 together with the corresponding average score*24. For any point in between, use interpolation to get the average score. Aiming point 41 is the centre of hole 5.

Aiming point	Average*24	Aiming point	Average*24
12	46	51	65
13	46	52	64
14	46	53	63
15	39	54	62
16	38	55	57
17	37	56	52
18	36	57	52
19	42	58	52
20	48	59	52
21	48	60	52
22	48	61	53
23	48	62	54
24	48	63	47
25	46	64	40
26	44	65	33
27	47	66	34
28	50	67	35
29	53	68	36
30	56	69	37
31	59	70	38
32	62	71	38
33	60	72	34
34	60	73	30
35	60	74	26
36	60	75	22
37	60	76	18
38	60	77	17
39	68	78	16
40	70	79	19
41	<u>72</u>	80	22
42	66	81	25
43	60	82	28
44	54	83	31
45	54	84	34
46	54	85	33
47	54	86	32
48	58		
49	62		
50	66		

Answer: 211

Let u be the object distance (ab), v be the image distance (cd) and k (ef) be the focal length. Then $1/u + 1/v = 1/k$ giving $k = uv/(u+v)$. Thus the denominator must contain the factors of u and v, adding to $u + v$ or a multiple thereof if there is to be any chance of a whole-number answer. As $u+v+k$ is odd (prime), we need two even numbers plus an odd number or three odd numbers. If all three are odd, $(u+v)$ is even, uv is odd, but $k(u+v)$ = uv, so that is out. So we need two even numbers (will be u and k because u and k have the same parity) and an odd number (v).

Now considering the extreme values of k. At the lower end, $13 \times 24/(13 + 24) = 312/37$ which is less than 10 so 12 must be the minimum value. At the other end, $96 \times 87/(96+87)$ = $8352/183$ = less than 46 so 45 is the maximum. As k is even, we are left with 12 14 16 18 24 26 28 32 34 36 38 and 42.

We can now list possible combinations of b, d and f, the consequent value of the last digit of k, and the sum, the last digit of which must be 1, 3, 7 or 9 since it is prime.

b	2	8	2	4	6	4	6	8
d	1	1	9	3	3	7	7	9
f	4	2	8	6	2	8	4	6
Sum	7	1	9	3	1	9	7	3

As b is an exact multiple of d, only the columns 1, 2 and 5 qualify. We now summarise possible candidates (remembering that u must be greater than k since $1/u < 1/k$):

Column	k	u	v	sum
1	34	52	n/i	
		62	n/i	
		72	n/i	
		82	n/i	
		92	n/i	
2	32	48	96	178 out
		58	n/i	
		68	n/i	
		78	n/i	
		98	n/i	
	42	58	n/i	
		68	n/i	
		78	91	211
		98	n/i	
5	12	36	18	66 out
		46	n/i	
		56	n/i	
		76	n/i	
		96	n/i	

Column	k	u	v	sum
	42	56	>100	
		76	n/i	
		86	n/i	
		96	n/i	

Only 211 is prime so the answer is 211.

Answer: 84588800688

dYSCALCULIA divides exactly by 9x8x7x6=16x9x7x3

Two digits ok; no segments always 'on'; some segments in any faulty digit always 'off'

A can only be 8 (bottom segment 'off')', but value divides by 16, so **ULIA** must divide by 16 (so 8 and 4). So **I** must be even (not=1) and can be 0, 4 or 8 (not 2 or 6 cf. right segments)

Thus the two undamaged digits must be those showing '4' and '5'.

d can only be 8, so, with above, the value is 845 **C** 8 **LCUL** I 8

L (must be even from **IA** above) can be 0 or 6 (in **6** or **b** form) or 8

U can be 0 or 8 (so end triplet **LI** 8 divisible by 16 – so 608, 688, 048, 848)

C can be 0, 6 (in **b** form) or 8

'dYSCALCULIA ' is also divisible by 9 – so digit sum must be multiple of 9.

So 'dYSCALCULIA' options are from

dYS	C	A	L	C	U	L	I	A
845	[0; 6; 8]	8	[0; 6; 8]	[0; 6; 8]	[0; 8]	[0; 8]	4	8

So 8+4+5+8+4+8=37+X=9N, but X=sum from even options shown, so is even <=40

845	[0; 6; 8]	8	[0; 6; 8]	[0; 6; 8]	[0; 8]	6	[0; 8]	8

So 8+4+5+8+6+8=39+Z=9M, but Z=sum from even options shown, so is even <=40

So X=8 or 26; Z=6 or 24 (digit sum=37+8=39+6=45 or 37+26=39+24=63)

X=8 options	X=26 options	Z=6 options	Z=24 options	
84588000048	84568668048	84568000608	84508088688	84588080688
84508800048	84568660848 *	84508600608	84508808688	84588088608
84508080048		84508060608	84508880688	84588800688 *
84508008048			84508888608	84588808608
84508000848			84588008688	84588880608

* only **84568660848** and **84588800688** divide exactly by 9x8x7x6=3024.

84568660848/3024=27965827

'CALCULIA'=68660848, so 27965827 shows each 7 as '7' (others 'non-numeral' patterns)

84588800688/3024=27972487

'CALCULIA'=88800688, so 27972487 shows leftmost 7 and rightmost 7 as '7' (middle 7 would not display right-hand segments); and 8 as '1' (others 'non-numeral' patterns)

Answer: £1,764

Formula for the sum of the series 1, 2, 3 · · · · · · n

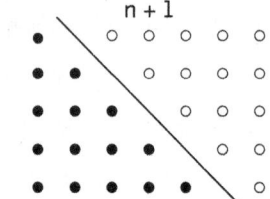

The sum of the sequence of numbers
1 + 2 + 3 + 4 · · · · · · + n can be seen from
the diagram to be equal to $n(n+1)/2$ (i)

Formula for calculating the sum of 'n' squares

Using the formula $(x+1)^3 = x^3 + 3x^2 + 3x + 1$ and substituting for various values of x gives -

$1^3 =$	$(0+1)^3 =$	0^3	$+ 3(0)^2$	$+ 3(0)$	$+ 1$
$2^3 =$	$(1+1)^3 =$	1^3	$+ 3(1)^2$	$+ 3(1)$	$+ 1$
$3^3 =$	$(2+1)^3 =$	2^3	$+ 3(2)^2$	$+ 3(2)$	$+ 1$
etc				
$n^3 =$	$(n-1+1)^3 =$	$(n-1)^3$	$+ 3(n-1)^2$	$+ 3(n-1)$	$+ 1$
$(n+1)^3 =$	$(n+1)^3 =$	n^3	$+ 3n^2$	$+ 3n$	$+ 1$

Adding the columns gives –

$$\sum_1^{n+1} n^3 \quad = \quad \sum_0^n n^3 \; + \; 3\sum_0^n n^2 \; + \; 3\sum_0^n n \; + \; (n+1)$$

Simplifying and substituting $n(n+1)/2$ for $\sum_0^n n$ from equation (i) gives - $(n + 1)^3 =$
3 Sum(n^2) + $3n(n+1)/2$ + n + 1

simplifying and factorizing gives – Sum(n^2) = $n(n+1)(2n+1)/6$ (ii)

Let n be the number of birthdays when so such presents were received, so from the Teaser
Sum(2n) = 1.15{Sum(2n) – Sum(n)}
Simplified this becomes – 23Sum(n) = 3Sum(2n)
Substituting in equation (ii) gives – 23{n(n+1)(2n+1)}/6 = 3{2n(2n+1)(4n+1)}/6
Which simplifies to $2n^2$ - 33n – 17 = 0
Using the quadratic formula this gives values of n of 17 and minus 0.5.

The number of birthdays for which no such presents were given is therefore 17 and from
equation (ii) these would have totaled £1,785.

Let 't' be the age at which the last birthday payment is made and Sum(t) be the total amount
paid up to and including this year if the payments had been made on every birthday.
Therefore Sum(t) = 1.075(S(t) – 1785) so Sum(t) = 25,585
From equation (ii) Sum(t) = t(t+1)(2t+1)/6 = 25,585
So t(t+1)(2t+1) = 25,585 x 6 = 153,510
The factors of 153,510 are 2, 3, 5, 7, 17 and 43, from which it can be seen that t = 42
satisfies the equation.

The final birthday present is therefore 42^2 = £1,764

76 FAMILY BUSINESS

Answer: 2385

We first write the three equations: abcd + acdb = dabc (1)
That proves that a cannot be more than 4 and d must be at least 2.
acdb + dbca = cabd (2)
That proves initially that c must be at least 3
bacd + dabc = cdab (3)
As a and b are different, c is now to be proves to be at least 4
Translating that into real numbers, we have
1000a + 100b + 10c + d + 1000a + 100c + 10d +b = 1000d + 100a +10b + d (1)
1000a + 100c + 10d + b + 1000d + 100b + 10c +a = 1000c + 100a +10b + d (2)
1000b + 100a + 10c + d + 1000d + 100a + 10b + c = 1000c + 100d + 10a +b (3)
Simplifying, we have 1900a + 91b + 109c = 989d (1)
901a + 91b + 1009d = 890c (2)
190a + 1009b + 901d = 989c (3)
We multiply (3) by 10 as a precursor to eliminating a
1900a + 10090b + 9010d = 9890c (3)
Subtracting and rearranging gives 9999b + 9999d = 9999c so that b + d = c (4)
Subtracting (2) from (1) to eliminate b:
999a + 109c – 1009d = 989d – 890c
Rearranging gives 999a + 999c = 1998d so that a + c = 2d (5)
Combining (4) and (5) a + b + d = 2d or a + b = d
So we have the following possibilities for a b c d

a	1	1	1	2	2	3	3	4	4
b	2	3	4	1	3	1	2	1	2
c	5	7	9	5	8	7	7	6	8
d	3	4	5	3	5	4	5	5	6

Starting with a = 1 we have the following possibilities, initially testing on equation (1):

1253 + 1532 = 2785 out
1374 + 1743 = 3117 out
1495 + 1954 = 3449 out

For a = 2
2153 + 2531 = 4684 out
2385 + 2853 = 5238 still alive

For a = 3
3174 + 3741 = 6915 out
3275 + 3752 = 7027 out

For a = 4
4165 + 4651 = 8816 out
4286 + 4862 = 9148 out

So there is one candidate and it survives all tests:

3285 + 5238 = 8523 and 2853 + 5382 = 8235

So Andrea is on 2385

Answer: 3 quid and 49 quid

Denominations X and Y with $1 \leq X < Y$, and X,Y having no common factor >1 (for if they are both divisible by N then all the totals made with them will be divisible by N).

A little trial-and-error with cases shows that it is impossible to make a total of

$(X - 1)(Y - 1) - 1$, but all totals from $(X - 1)(Y - 1)$ upwards are possible. So we need

$$(X - 1)(Y - 1) \leq 100.......(1).$$

In particular $X \leq 10$.

If X or Y equals 1, 2, 4, 5 or 10 then 100 would not need a mixture of denominations to make 100. So X is one of 3, 6, 7, 8 or 9.

It is clear that all totals from $XY + (X - 1)(Y - 1)$ upwards can be made in two different ways (since the XY can be X @ Y or Y @ X). So we need

$$XY + (X - 1)(Y - 1) > 230......(2)$$

X=3?

From (1) $Y \leq 51$ and from (2) $Y \geq 47$. To have no common factor with X, and for 100 not to be possible with Ys alone, Y must be 47 or 49. But

230 = 61@3 + 1@47 = 14@3 + 4@47
   = 44@3 + 2@49 only

So X=3 and Y=49 work.

X=6?

From (1) $Y \leq 21$ and from (2) $Y \geq 22$! Any higher X fails similarly.

Answer: 13, 17, 29 and 31

Hypotenuse value (3-figure<1000) is mean of two different odd squares.
$47^2=2209$, so both odd numbers are primes P, Q<47.

Find $(P^2+Q^2)/2$= 3-fig. palindromic value, repeated, from valid distinct pairs selected from 3, 7, 11, 13, 17, 19, 23, 29, 31, 37, 41, 43

	P=	3	7	11	13	17	19	23	29
	P^2=	9	49	121	169	289	361	529	841
Q	Q^2								
13	169	<100	109	145					
17	289	149	169	205	229	Repeats (from below)			
19	361	185	205	241	265	325			
23	529	269	289	325	349	409	445		
29	841	425	445	481	505	565	601	685	
31	961	485	505	541	565	625	661	745	901
37	1369	689	709	745	769	829	865	949	
41	1681	845	865	901	925	985			
43	1849	929	949	985			>999		

So repeated hypotenuse value H=505, 565, 929 or 949

$H^2=A^2+B^2$, with A, B integers approx. equal if triangle is close to a 45° triangle.

Approximate A, B by setting A=B so A $\approx$ H/$\sqrt2$ and with <2% diff. between A and B

H	H/$\sqrt2$	H/$\sqrt2$]-1%	Test A=	for B=$\sqrt{(H^2-A^2)}$ integer
505	357.1	353.5	353 to 358	no solution
565	**399.5**	**395.5**	**395 to 400**	**A=396, B=403**
929	656.9	650.3	650 to 657	no solution
949	671.0	664.3	664 to 672	no solution

So 'near-45° triangle' has sides 396, 403, 565

The required repeated hypotenuse H=565 arises from prime pairs [13, 31] and [17, 29]
$565=[31^2+13^2]/2=[961+169]/2=[29^2+17^2]/2=[841+289]/2=1130/2$ from table above

So sisters/brothers age pairings are [13 with 31] and [17 with 29] (or vice versa)

In ascending order of age 13, 17, 29 and 31

Answer: 2,601

Let the original four-digit number be "abcd" and the five-digit number be "efghi"

$(abcd)_{base10} \times 4 = (efghi)_{base10}$, and $\qquad\qquad$ (i)
$(efghi)_{base7} = (abcd)_{base10}$
So $(efghi)_{base7} \times 4 = (efghi)_{base10}$
$4 \times (7^4e + 7^3f + 7^2g + 7h + i) = 10000e + 1000f + 100g + 10h + i$

This simplifies to: $31f + 8g = \tfrac{1}{4}(132e - 6h - i)$ $\qquad\qquad$ (ii)

As "efghi" is base 7 its digits' maximum values are 6. Also, for a four-digit number to produce a five-digit number when multiplied by 4 the maximum "e" can be is 3. "e" is therefore 1,2 or 3.

"abcd" is a perfect square so its last digit "d" can only be 0,1,5,6 or 9. When multiplied by 4 these can only become 0, 4, or 6 so "i" is restricted to these numbers.

As "efghi" is divisible by 4 so must its last two digits "hi" be divisible by four. Using such values of "hi" and the possible values of "e" the following table shows, from (ii), the resulting values of $(31f + 8g)$.

| h | i | e | | | f | g |
(divisible by 4)		1	2	3	(corresponding to bold values)	
0	0	33	66	99		
0	4	**32**	65	98	0	4
1	6	30	**63**	96	1	4
2	0	30	**63**	96	1	4
2	4	29	**62**	95	2	0
3	6	27	60	**93**	3	0
4	0	27	60	**93**	3	0
4	4	26	59	92		
5	6	**24**	57	90	0	3
6	0	**24**	57	90	0	3
6	4	23	56	89		

The possible values of $(31f + 8g)$ can be seen from the table to be between 23 and 99. By inspection this is only possible if "f" is 0, 1, 2 or 3 (then when g is 0). From inspection only the values in bold in the table can be arrived at for values of "f" and "g" between 0 and 6 being substituted in $(31f + 8g)$. These values of "f" and "g" are shown in the end columns.

These give the following solutions –

e	f	g	h	i
1	0	4	0	4
2	1	4	1	6
2	1	4	2	0
2	2	0	2	4
3	3	0	3	6
3	3	0	4	0
1	0	3	5	6
1	0	3	6	0

This is the only perfect square

(Note that there are two other solutions with values of "i" of 2. These are not picked up because of the perfect square restriction.)

"efghi" is therefore 10,404 and from (i) above "abcd" = 10,404 /4 = 2,601

80 · WILLIAM'S PRIME

Answer: 4177123

Prime numbers of two or greater digits must end in 1, 3, 7 or 9. Therefore the L in WILL must be 1, 3, 7 or 9. The lowest valid prime value of WILL is 1277.

The number of digits in the product must be in the range 5 to 7, since 4 or below digits is less than the minimum possible and 8 or more is above the maximum possible (9811 * 8 *73 = 5729624).

If WILL is 1277, the lowest possible prime value of AM is 43.

There are multiple 6-digit products using the lowest value for WILL (e.g. 1277 * 2* 43 = 109822 and 1277 * 2 * 53 = 135362). Similarly, there are multiple 7-digit products using higher values of WILL (e.g. 9833 *8 *17 = 1337288 and 9833 * 8 *41 = 3225224). Therefore the number of digits in the product can be deduced as 5.

Performing the calculation of WILL * I * AM using 1277 * 2* 43 already gives a 6-digit total. The Solver will then realise that I must be 1 to give a 5-digit total. Then A is at least 2, so W can be no more than 4 to give a 5-digit product.

The possible values of WILL are therefore 2133, 2177, 2199, 3177, 3199, 4133, 4177, and 4199. After removing the non-prime numbers*, this reduces to 4133 and 4177.

Using these values for WILL, AM must be less than 29, since 4133 * 1 * 29 = 119857 has 6 digits. Therefore, AM (a prime number not containing 1) must be 23 and WILL must be 4177. WILL * I * AM is 96071 with 5 digits.

WILLIAM is therefore 4177123.

*Non-prime values of WILL:
2133 is divisible by 3
2177 is divisible by 7
2199 is divisible by 3
3177 is divisible by 3
3199 is divisible by 7
4199 is divisible by 13

Answer: '4'

100xSum(1/n) for n = 1 to 20 = 359.77

360-359.77=0.23, so '20' sector central angle =0.23+100x(1/20)=5.23 degrees

There are twenty [A, B, C] sets of adjacent sectors.
The angle sum X=100(1/A+1/B+1/C)=100(BC+AC+AB)/ABC (+0.23 if A, B or C=20)

Tabulate for angle sum X (to next higher integer) for main elimination

A	B	C	Angle sum X	A	B	C	Angle sum X
1	18	4	131	19	7	16	26*
18	4	13	39**	7	16	8	34**
4	13	6	50	16	8	11	28*
13	6	10	35**	8	11	14	29*
6	10	15	34**	11	14	9	28*
10	15	2	67	14	9	12	27*
15	2	17	63	9	12	5	40***
2	17	3	90	12	5	20	34*
17	3	19	45**	5	20	1	126
3	19	7	53	20	1	18	111

We can eliminate many, simply, by A+B+C>X* and others by A+B+C~X and x3 for smallest A, B or C makes total score>=X** and one other for which any total score>X***

For valid sets tabulate 6 totals and eliminate if total score>X (s=single, d=double, t=treble)

sdt=A+2B+3C; std=A+3B+2C; dst=2A+B+3C; dts=2A+3B+C; tsd=3A+B+2C; tds=3A+2B+C

A	B	C	sdt	std	dst	dts	tsd	tds	X
1	18	4	**49**	**63**	**32**	**60**	**29**	**43**	131
4	13	6	**48**	>X	**39**	>X	**37**	**44**	50
10	15	2	**46**	**59**	**41**	>X	**49**	**62**	67
15	2	17	>X	**55**	>X	**53**	>X	>X	63
2	17	3	**45**	**59**	**30**	**58**	**29**	**43**	90
3	19	7	>X	>X	**46**	>X	**42**	>X	53
5	20	1	**48**	**67**	**33**	**71**	**37**	**56**	126
20	1	18	**76**	**59**	**95**	**61**	**97**	**80**	111

From these valid scores only two six-value consecutive sequences arise

41+42+43+44+45+46=261 leaving 501-261=240 (no 3-dart out shot) and

58+59+60+61+62+63=363 leaving 501-363=138 (e.g. [t20, t20, s18 or d9 or t6], etc.)

58, 60, 61, 62, 63 unique entries, but 59 score could be from 3 options, each allied to a sector triplet for a unique entry. From these tally s, d, t zone hits for the common sector set

1-sdt	2-sdd	3-s	4-sd	10-st	15-dt	17-t	18-stt	20-d
1-sdt	2-ssd	3-sd	4-sd	10-t	15-d	17-tt	18-stt	20-d
1-sdtt	2-sd	3-s	4-sd	10-t	15-d	17-t	18-sdtt	20-sd

Only '4' sector certainly hit just twice – s4 and d4

Answer: 11 hectares

There are five ways of planting the first two hedges but only one of which divides it into three. There is then one way of planting the other two hedges that divides the rectangular field into nine smaller fields.

If the rectangular field has width 2x and length 2y and the smaller fields have areas Δ_1, Δ_2, etc. the fields appear as shown in the diagram below.

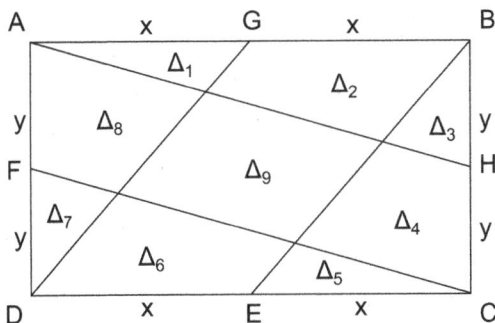

It can be seen by inspection that $\Delta_1 = \Delta_5$, $\Delta_3 = \Delta_7$, $\Delta_2 = \Delta_6$ and $\Delta_4 = \Delta_8$ because DG is parallel to EB and AH is parallel to FC. Triangles AGD, ABH, BEC and DFC all have area xy so $\Delta_7 + \Delta_8 + \Delta_1 = \Delta_1 + \Delta_2 + \Delta_3 = \Delta_3 + \Delta_4 + \Delta_5 = \Delta_5 + \Delta_6 + \Delta_7$. Substituting for Δ_5 and Δ_7 gives $\Delta_3 + \Delta_8 + \Delta_1 = \Delta_1 + \Delta_2 + \Delta_3 = \Delta_3 + \Delta_4 + \Delta_1 = \Delta_1 + \Delta_6 + \Delta_3$ hence $\Delta_8 = \Delta_2 = \Delta_4 = \Delta_6$ and the diagram can be redrawn as follows:-

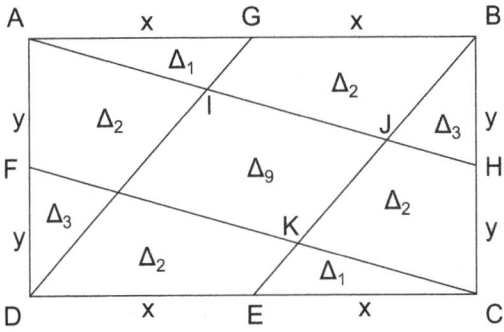

Triangle ABJ is similar to triangle AGI and its sides are twice as long so its area is four times as large, therefore $\Delta_1 + \Delta_2 = 4\Delta_1$ or $\Delta_2 = 3\Delta_1$. Triangle BKC is similar to triangle BJH and its sides are twice as long so its area is four times as large, therefore $\Delta_3 + \Delta_2 = 4\Delta_3$ or $\Delta_2 = 3\Delta_3$. Hence $\Delta_1 = \Delta_3$.

The triangle AGD has area xy $= \Delta_1 + \Delta_2 + \Delta_3 = \Delta_1 + 3\Delta_1 + \Delta_1 = 5\Delta_1$ so $\Delta_1 = xy/5$ and $\Delta_2 = 3\Delta_1 = 3xy/5$. The rectangle ABCD has area 4xy $= 2\Delta_1 + 4\Delta_2 + 2\Delta_3 + \Delta_9 = 2\Delta_1 + 12\Delta_1 + 2\Delta_1 + \Delta_9 = 16\Delta_1 + \Delta_9$ so $\Delta_9 = 4xy - 16\Delta_1 = 4xy/5$.

The area of the largest field (Δ_9) is one fifth of the area of the rectangular field or 11 hectares.

83 SPORTY SET

Answer: 65

Let the numbers be T, B, S and Q.

Since Q is the least, that is the maximum number who could play all four sports.

To find the minimum number who could play all four sports we must maximise the number who avoid at least one. That maximum will occur when the 100-T non-tennis players, the 100-B non-badminton players, the 100-S non-squash players, and the 100-Q non-table-tennis players are all different people (if possible). The maximum number avoiding a sport will then be
(100 - T) + (100 - B) + (100 - S) + (100 - Q) = 400 - (T + B + S + Q)
(or 100 if that total exceeds 100).

So in this case the minimum number playing all four sports is (T + B + S + Q) − 300.

For this to be a 2-figure number T + B + S + Q must be at least 310. The fact that they use different non-zero digits means that these four numbers must (in some order) be 9-, 8-, 7- and 6- with the four unit digits adding to 10 or more. The units can add to at most 5+4+3+2=14. We tabulate the possibilities:

T+B+S+Q	Minimum M	Q=6- (it must be a multiple of M)
310	10	Q=60, but 0 not allowed
311	11	Q=66, but repeated digit not allowed
312	12	Q=60, but 0 not allowed
313	13	**Q=65, O.K.**
314	14	No such Q exists

[e.g. T=91, B=83, S=74 and Q=65]

84   TIMELY COINCIDENCE

Answer: 19/05/61

The time clock over one day can display anything from 00 00 00 midnight to 23 59 59 one second before the next midnight. Thus there are 24 x 60 x 60 = 86,400 possible displays.

The date 'clock' can display 01 01 00 (1st. January at the beginning of a century) to 31 12 99 (New Year's Eve just before the beginning of the next century). To be accurate, we have to work over a 400 year period to allow for the fact the centenary years are not leap years unless divisible by 4. Thus, in a 400-year period, there are 303 non-leap years and 97 leap years so the total number of days in the period is 303 x 365 + 366 x 97 = 146,097. Thus there will be a total of 146097 x 86400 = 12622780800 possible displays over the 'quadrennium'.

If there is to be any chance of coincidence, the two clocks will have to show anything from 01 – 23, 01 – 12 and 00 – 59 i.e. 23 x 12 x 60 = 16,560 possible combinations over a century. Multiplied by 4 that gives 66240 over the period. Thus the probability of coincidence is 66240/12622780800 = 1 in 190561.3043 or 1 in just over 190561.

So the daughter was born on 19th May 1961.

85   TRIPLE JUMP

Answer: 1 2 6 8 2 5

In this proof the position of a card will be denoted by 'p' followed by its position number. For example p4 cannot be a 2 because this would require a 2 to be in p7 and we know that p7 is a 7.

Table 1 has been populated in accordance with the rules and shows all the valid positions for the cards.

Table 1

1	2	3	4	5	6	7	8	9	10	11	12	13	14	15	16	17	18	19	20	21	22	23	24
	2	2		2	2			2	2		2	2		2			2	2	2	2	2		2
3	3		3	3	3		3	3	3		3	3	3		3	3	3		3	3	3		3
4		4	4		4		4	4		4			4	4		4		4	4		4		4
	5			5	5	5		5		5	5	5		5		5	5	5		5		5	5
	6	6	6	6	6			6	6	6	6	6			6	6	6	6	6				6
						7							7									7	
8	8	8	8					8	8	8	8						8	8	8	8			
1	1		1							1	1		1							1	1		1

There must be an ace (1) in p1, p2 or p4 and there has to be an 8 in p1, p2, p3 or p4. My solution hinges on considering the following three cases.
Case 1: Ace is in p1 – This eliminates the 8's from p1, p2, p3, p10, p11, p12, p19, p20 and p21
Case 2: Ace is in p2 – This eliminates the 8's from p2, p3, p4, p11, p12, p13, p20, p21 and p22
Case 3: Ace is in p4 – This eliminates the 8's from p4, p13 and p22

Each of these cases lead to a definitive conclusion without recourse to trial and error. If Case 3 is followed and Table 1 is depopulated accordingly, we find that p24 = 1, p23 = 7 and P22 = 5. The discovery that p22= 5 leads rapidly to a conclusion that Case 3 does not provide a solution. Similarly, if Table 1 is depopulated by following Case 2 we find that p24 must be 2 or 3, p17 must be 2, 3 or 5 and p5 must be 2, 3 or 5. If p17 = 2 then p2 = 3 and p5 =5. If P17 = 3 then p24 = 2 and p5 = 5. If p17 = 5 then p5 = 5. Therefore p5 must = 5. When the table is depopulated from here it leads to p8 = 4 and p14 = 4, which cannot be true as this gap is not correct for 2 four's.

Following Case 1 leads to the depopulation of Table 1 and results in Table 2 below.

Table 2

1	2	3	4	5	6	7	8	9	10	11	12	13	14	15	16	17	18	19	20	21	22	23	24
	2	2		2	2		2	2			2		2			2			2				
	3			3	3		3	3	3		3		3		3	3	3		3				3
								4					4					4					4
	5			5			5				5		5			5			5				5
	6	6		6			6	6		6					6	6			6				6
						7							7									7	
		8										8								8			
1								1												1			

The following logic takes the solution beyond this point in Table 2.

p10 must equal 3 or 6, p16 must equal 3 or 6 and p5 must be 2, 3 or 6.

If p10 = 3 then p16 = 6 and if p10 = 6 then p16 = 3. Therefore p5 cannot be 3 or 6 and p5 must be equal to 2. p5 = 2 leads to p2= 2 and p8= 2 which provides further depopulation and rapidly leads to the full solution.

If we remove the restriction that p7 = 7, then there are four possible solutions (or two and two palindromes) as illustrated below. There is only one solution when the p7 = 7 restriction is applied.

8 2 1 6 2 7 4 2 5 8 6 4 1 7 5 3 4 6 8 3 5 7 1 3

3 7 8 1 3 4 5 6 3 7 4 8 5 1 6 4 2 7 5 2 8 6 2 1

1 2 6 8 2 5 7 2 4 6 1 5 8 4 7 3 6 5 4 3 1 8 7 3

3 1 7 5 3 8 6 4 3 5 7 1 4 6 8 5 2 4 7 2 6 1 2 8

Answer: 7250 seconds

Let A's speed be "a" and for simplification change the frame of reference by deducting A's speed from each of the three runners' speeds so that A remains at the starting position – point 0.

B's speed is now 1.42a – a = 0.42a. If after 4625s of running distances are 85m apart then

4625 x 0.42a/400 or 1942.5a/400 must have a remainder of 85, 170, 230 or 315.

From inspection "a" must be an even number and the following table shows the remainders for even single digit values of "a" –

	Value of "a"			
	2	4	6	8
Distance run – 1942.5a	3885	7770	11655	15540
Remainder, distance run divided by 400	285	**170**	55	340

It can be seen that "a" must be 4m/s as only this speed gives a valid remainder value.

The position of A after 4625s is 0, that of B is 170 so C's position must be 85.
Let C's excess speed relative to A be "c", where 0.42<c<1 since C's speed is less than twice that of A.
Therefore c x 4 x 4625/400 must have a remainder of 85.
This converts to 18500c = 400k + 85 which becomes 3700c – 80k = 17 for some whole number k.
Now let C = 100c, where 42<C<100, then 37C-80k = 17
37C must end in 7, so C ends in 1 and C = 51, 61, 71, 81 or 91.
The only value that gives a whole number for k is C=61.

Let the total running time for the runners to be 90m apart be "t".
At time t A's position is 0, B's position is the remainder from 0.42x4xt/400 and C's position the remainder from 0.61x4xt/400.
The allowable remainder values are 90, 180, 220 and 310, so
For B, 1.68t = 400y + 90, 180, 220 or 310, for some whole number y, and correspondingly
For C, 2.44t = 400z + 180|310, 90, 310 or 90|220 (the order of the runners, clockwise, being ABC|CAB, ACB, BCA or BAC|CBA) for some whole number z

To eliminate "t" in the above, multiply the first equation by 61 and the second by 42 to get
102.48t = 24400y + 61(90, 180, 220 or 310) = 16800z + 42(180|310, 90, 310 or 90|220)

The first and fourth possibilities don't work, as 61x90 and 61x310 aren't divisible by 20 and all other terms are. We are left with:

2440y + (1098 or 1342) = 1680z + (378 or 1302), which reduce to:
(1) 61y + 18 = 42z
(2) 61y + 1 = 42z

The first solution of (1) is y=30, z=44 and t = (400y+180)/1.68 = 7250
The first solution of (2) is y=11, z=16 and t = (400y+220)/1.68 = 2750
The next solution of (2) is y=53, z=77 and t = (400y+220)/1.68 = 12750

Therefore the first time after 4625s that there will be gaps of 90m is at a time of 7250s

Answer: 1785

Possible PARs [Increasing order]	Possible larger PARs
1236	-
1248	-
1284	-
1296	-
1326	-
1352	4896
1365	2987
1378	2496, 4692
1428	-
1456	2987, 3978✓
1498	3672
1632	-
1648	-
1734	2856, 2958✓
1768	-
1785	2346, 2369, 2496✓, 3264, 3296, 4692✓
1836	2754✓
1854	2369, 3296, 3672✓
1872	-
1938	2754✓
1957	2346, 3264, 3468, 4386

None of the remaining possibles 1976, 2163, 2184, 2346, 2369, 2496, 2678, 2754, 2781, 2856, 2958, 2987, 3162, 3264, 3296, 3468, 3672, 3876, 3978, 4182, 4386, 4692, 4896 have associated possible larger PARs. Of the above possibilities, those with a ✓ give possible PARTYs:

1456 and 3978, missing digit 2 a common factor
1734 and 2958, missing digit 6 a common factor
1785 and 2496, missing digit 3 a common factor ✓✓
1785 and 4692, missing digit 3 a common factor ✓✓
1836 and 2754, missing digit 9 a common factor
1854 and 3672, missing digit 9 a common factor
1938 and 2754, missing digit 6 a common factor

So my PAR was 1785 and Sam's and Beth's PARs were 2496 and 4692 in either order.

Answer 10

In the table below the outcome frequency of possible scores is given for 1, 2 and 3 dice (e.g. there are 21 ways of obtaining a score of 13 with 3 dice from a possible 216 outcomes).

This is followed by the frequency of beating that score with that number of dice (e.g. the number of ways of beating 13 with three dice is the sum of the ways of getting 14, 15, 16, 17 and 18, i.e. 15+10+6+3+1 = 35).

Score	1	2	3	4	5	6	7	8	9	10	11	12	13	14	15	16	17	18
1 die	1	1	1	1	1	1												
	5	4	3	2	1	0												
2 dice		1	2	3	4	5	6	5	4	3	2	1						
		36	35	33	30	26	21	15	10	6	3	1	0					
3 dice				1	3	6	10	15	21	25	27	27	25	21	15	10	6	3	1
			216	216	215	212	206	196	186	160	135	108	81	56	35	20	10	4	1	0

The probability of Liam winning is given by the frequency in the table divided by 6, 36 or 216 as appropriate (e.g. if Callum scores 13, Liam's chances with three dice are 35/216).

Only in the case of a score of 10 by Callum with three dice are the conditions of the puzzle met.

Liam can beat him with two dice (with two 6s or a 5 and a 6), and if he can use three dice, his chances increase from 3 in 36 to 108 in 216, i.e. from 1 in 12 to 6 in 12, so a whole number of times greater.

Answer: 4

With each possible score – 1, 2, 3, 4, 5 and 6 uppermost the 7th score must repeat one of these.

Each 7-digit number has just one single-figure prime factor – different for each. So the repeated score can't be 3 or 6 – otherwise digit sum div. by 3 and so for each permed number.

So initial possibilities are:-

SMALLEST	LARGEST	ROYGBIV
1123456	6543211	
[factor 2 – not 3, 5 or 7]	[not factors 2,3,5 or 7]	n/a
1223456	6543221	
[factor 2 – not 3, 5 or 7]	[not factors 2,3,5 or 7]	n/a
1234456	6544321	
[factor 2 – not 3, 5 or 7]	[factor 7- not 2, 3 or 5]	see below
1234556	6554321	
[factor 2 – not 3, 5 or 7]	[not factors 2,3,5 or 7]	n/a

Central 4 in 1234456=2xN and 6544321=7xM also in ROYGBIV (so G=4). ROY4BIV must also be div. by 5, so V=5 and rainbow value is ROY4BI5 with R, O, Y, B and I not in same digit positions as 1234456 and 6544321.

[R][O][Y]4[B][I]5 perms from [2,3,4][1,3,4,6][1,2,6]4[1,2,6][1,3,4,6]5 with clear restrictions.

2164??5 and 2614??5 – invalid – impossible to perm from above

2314645=5x.. – valid, but 2364145=5x7x.. – invalid – unambiguous given R and O

24[1,6]4[1,6]35 – ambiguous perms – 2414635=5x.. and 2464135=5x.. – no factor 7

3124645=5x.. – valid, but 3164245=5x7x.. – invalid – unambiguous given R and O

34[1,2,6]4[1,2,6][1,6]5 – ambiguous perms – 3414265=5x.. and 3424165=5x.. – no factor 7
– 3424615=5x.. and 3464215=5x.. – no factor 7

3614245=5x.. – valid, but 3624145=5x7x.. – invalid – unambiguous given R and O

41[2,6]4[2,6]35 – ambiguous perms – 4124635=5x.. and 4164235=5x.. – no factor 7

43[1,2,6]4[1,2,6][1,6]5 – ambiguous perms – 4314265=5x.. and 4324165=5x.. – no factor 7
– 4324615=5x.. and 4364215=5x.. – no factor 7

46[1,2]4[1,2]35 – ambiguous perms – 4614235=5x.. and 4624135=5x.. – no factor 7

For each unambiguous case above, I(ndigo die)=4

Answer: 29

If my number is N, then one of the square jigsaws is at least NxN (to accommodate the 1xN piece). Also, to use at least two pieces, a 1x1 and 2x2 jigsaw are impossible. Furthermore, we can soon see that there are not enough small pieces to make two separate 3x3 jigsaws. Therefore the total minimum area of the rectangles must be at least $N^2 + 9 + 16$.

If N=15 a quick count gives a total area of the pieces as 210, way short of $15^2 + 25$. For subsequent N we calculate the areas cumulatively below:

Number N	Rectangles of area =N	Area of those rectangles	Total area T of all rectangles	T-N²≥25?
16	3	48	258	
17	1	17	275	
18	3	54	329	
19	1	19	348	
20	3	60	408	
21	2	42	450	
22	2	44	494	
23	1	23	517	
24	4	96	613	37
25	2	50	663	38
26	2	52	715	39
27	2	54	769	40
28	3	84	853	69
29	1	29	882	41
30	4	120	>999	

In no case is $T≥(N+1)^2 + 25$ and so the jigsaws are NxN and at least two others totalling $T - N^2$ in area. Of those numbers listed in the right-hand column above, only 41 can be expressed as a sum of some of 9, 16s, 25s, 36, 49 and 64. So the only possibility is N=29 and all the pieces in this case *can* be used to make square jigsaws of sides 29, 5 and 4.

[For completeness, one such possible layout can be seen on the next page.]

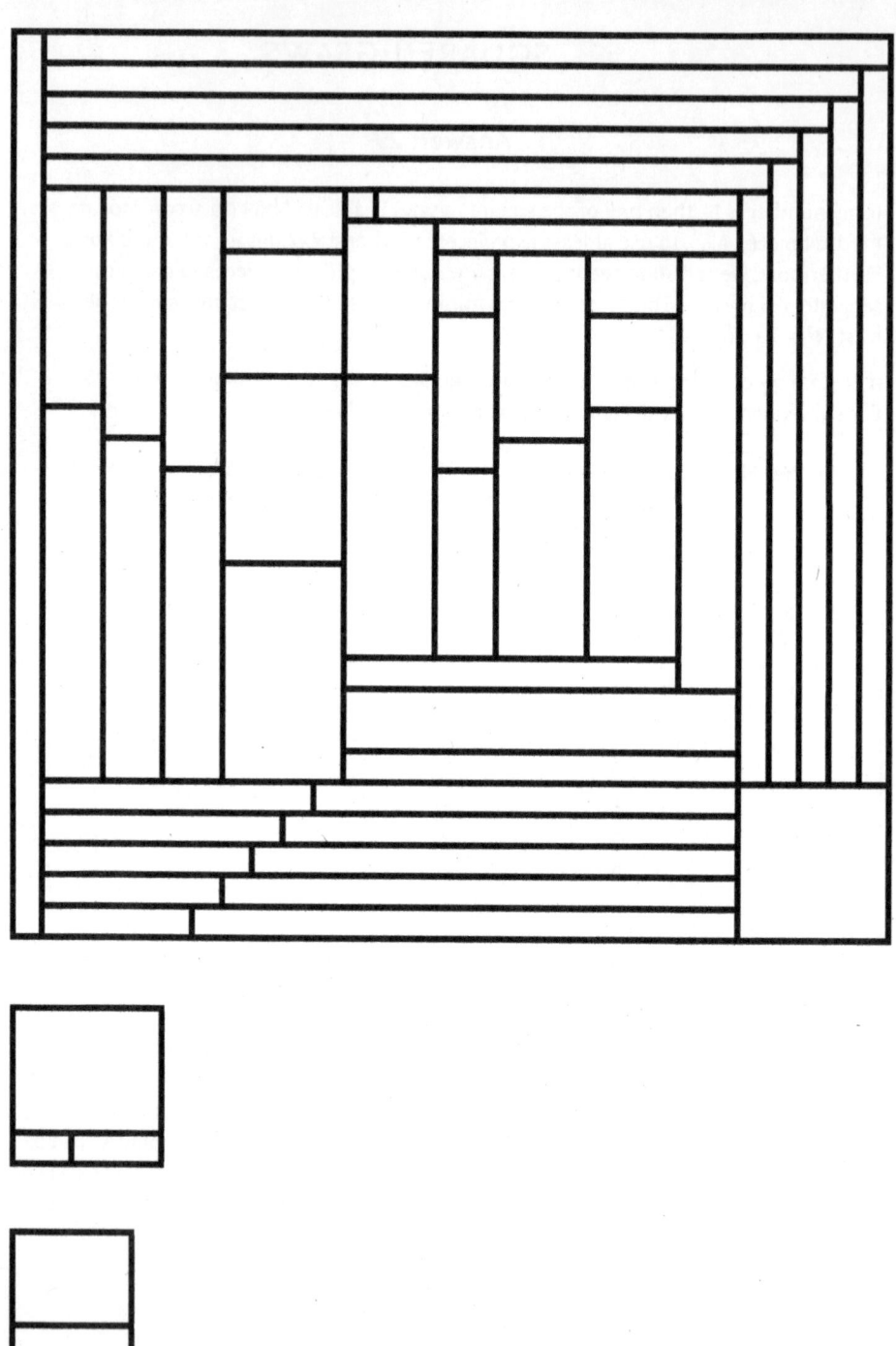

91 POT SUCCESS

Answer: 13

At the point in the match where my PS is a whole number, let a be the number of pots and b be the number of attempts up to that point, so $100a/b$ is a whole number. After then potting n balls, my PS has increased to $100(a+n)/(b+n)$, and after a further n pots it is $100(a+2n)/(b+2n)$. After the final miss, the PS is $100(a+2n)/(b+2n+1)$. Thus, all of the following numbers are whole:

$100a/b$, $100(a+n)/(b+n)$, $100(a+2n)/(b+2n)$ and $100(a+2n)/(b+2n+1)$

Now let $x = 100(a+2n)/(b+2n)$ and $y = 100(a+2n)/(b+2n+1)$, so $x/y = (b+2n+1)/(b+2n)$, or $y = (x-y)(b+2n)$

I potted "several" balls in a row, so n must be at least, say, 3, and $b+2n+1$ is at least 8.

Then we have $x/y = 8/7$, $9/8$, $10/9$, ... etc. For the first of these, we could have $(x,y) = (8,7)$, $(18,14)$, ... $(96, 84)$. x and y are PS values so can't be more than 100, so x-y is no more than 12. Subsequent values of x/y will not allow x-y to be any more than 12. Try x-y = 1, 2, 3 ... in turn:

<u>x-y = 1</u>, so $y = b+2n$
$100(a+2n) = x(b+2n) = y(y+1)$, so $y(y+1)/100$ is whole. The only possibilities for y are 24 and 75.
If y=24, a+2n=6, but n is 3 or more, so that doesn't work.
If y=75, a+2n=57 and b+2n=75, so b-a=18, both a and b are odd, and 100a/b is whole, so b is divisible by 5. The only possibilities that work are (a,b)=(7,25) and (27,45).

<u>x-y = 2</u>, so $y = 2(b+2n)$ and $a+2n = y(y+2)/200$ which must be whole. The only possibilities for y are 48 and 50.
If y=48, a+2n=12 and b+2n=24, so b-a=12, both a and b are even and 100a/b is whole. The only possibility is (a,b)=(4,16).
If y=50, a+2n=13 and b+2n=25, so b-a=12, both a and b are odd and 100a/b is whole. The only possibility is (a,b)=(3,15).

<u>x-y = 3</u>, so $y = 3(b+2n)$ and $y(y+3)/300$ is whole. The only possibility is y=72 and (a,b)=(0,6) – not allowed.

<u>x-y = 4, 6, 7, 8, 9, 11 and 12</u> give no solutions. For example, if x-y=7, y=7(b+2n) and $y(y+7)/700$ is whole, so y is a multiple of 7 and y or y+7 must be a multiple of 25, which doesn't work.

<u>x-y = 5</u>, so $y = 5(b+2n)$ and $y(y+5)/500$ is whole. The only possibility is y=75. Then a+2n=12 and b+2n=15, b is odd and 100a/b is whole. The only possibility is (a,b)=(2,5)

<u>x-y = 10</u>, so $y(y+10)/1000$ is whole. The only possibility is y=50, but a+2n=3, which is impossible

Summarising the various possibilities with the PS values at the various stages:

a	b	n	PS	PS	PS	PS	Balls potted	Balls attempted
7	25	25	28	64	76	75	57	76
27	45	15	60	70	76	75	57	76
4	16	4	25	40	50	48	12	25
3	15	5	20	40	52	50	13	26
2	5	5	40	70	80	75	12	16

Only if I potted 13 balls can you work out all the PS values.

Answer: 530,864,197

Let the correct number be "ab · · · · · ·" depending on the number of digits.

i)   For two digits – ab = ba x 43/25 = 1.72ba
for "b" to be an integer, and as "a" can't be a 0, then "a" must be 5
so 5b = 1.72 x b5 : 50 + b = 1.72(10b + 5) = 17.2b + 8.6 : 16.2b = 41.4 and
multiplying by 5 -
81b = 207, as 81 doesn't divide exactly into 207, "b" can't be an integer so this can't be
a solution.

ii) For three digits – abc = 1.72 bca, and again "a" must be 5.
So 500 + 10b + c = 1.72(100b + 10c + 5) : 162b + 16.2c = 491.4   multiplying by 5 –
810b + 81c = 2,457, again 81 doesn't divide exactly into 2,457 so three digits doesn't
have a solution.

iii) For four digits – 8,100b + 810c + 81d = 24,957 – no solution

iv) For five digits – 81,000b + 8,100c + 810d + 81e = 249,957 – no solution

A pattern is emerging as shown in the following table-

Number of Digits	RHS Value
2	207
3	2,457
4	24,957
5	249,957
6	2,499,957
7	24,999,957
8	249,999,957
9	2,499,999,957
10	24,999,999,957

The number of digits in the RHS value being tested for divisibility by 81 is too many for
most calculators so a short cut non-calculator test is applied in the following table. To test
for divisibility by 81 the last digit of the number is removed and from the remaining
number, 8 times the removed last digit is deducted. If after repeated application exactly
0 is left then the number is divisible by 81.

Number of Digits								
2	3	4	5	6	7	8	9	10
207	2,457	24,957	249,957	2,499,957	24,999,957	249,999,957	2,499,999,957	24,999,999,957
-36	189	2,439	24,939	249,939	2,499,939	24,999,939	249,999,939	2,499,999,939
	-54	171	2,421	24,921	249,921	2,499,921	24,999,921	249,999,921
		9	234	2,484	24,984	249,984	2,499,984	24,999,984
			-9	216	2,466	24,966	249,966	2,499,966
				-27	198	2,448	24,948	249,948
					-45	180	2,430	24,930
						18	243	2,493
						-63	0	225
								-18

Only for 9 digits is the RHS value exactly divisible by 81 giving a quotient of 30,864,197.

So $10^7b + 10^6c + 10^5d + 10^4e + 10^3f + 10^2g + 10h + i = 30,864,197$

As the first digit of the correct number, "a" is 5, the correct number is 530,864,197.

Answer 56

Very little calculation is necessary if three points are appreciated-

Evidently either an odd or even score is unique but not both.

Trebling a number does not change its odd or even character.

The unique answer is far more likely to be even as the arrangement of numbers on a dartboard give far fewer even 'adjacent triads'.

(20) 1 18 4 13 6 10 15 <u>2</u> <u>17</u> 3 19 <u>7</u> 16 8 11 <u>14</u> 9 <u>12</u> 5 <u>20</u> (1)

Central number	Number to be trebled		
	first	central	third
7	80	56	74
14	56	62	52
12	44	50	36
20	36	66	28
2	64	38	68
17	26	56	28

Only <u>56</u> appears three times.

For the sake of completeness the table is continued below –

1	79	41	75
18	25	59	31
4	71	43	61
13	31	49	35
6	55	41	49
10	43	51	61
15	47	57	31
3	73	45	77
19	35	67	43
16	45	63	47
8	67	51	57
11	49	55	61
9	63	53	59
5	61	47	77

43, 47 and 61 are possible odd scores (31 not allowed because central numbers of 18 and 13 are too close to allow nine different numbers, also 49 is not allowed for the same reason); confirming 56 as the correct answer.

Answer: (a) 7 (b) 10

First observe that each circuit is a simple loop, like a necklace of lights. If there are n lights and switches in a circuit, label them as follows: Choose a random light as number 1, and choose 2 to be one of the lights toggled by switch 1. Since 2 toggles 1, choose 3 to be the other light toggled by 2, 4 the other light toggled by 3, and so on. Eventually we will reach the light labeled n, and this must toggle light 1 in order to satisfy the condition that each switch toggles exactly two other lights in addition to its own.

Also note that starting with all lights off, activating each switch once will turn all the lights on. This is because each light gets toggled three times.

We will now show that any combination of n lights (including single lights) can be switched on if and only if n is not a multiple of 3. Note that the number of light states, 2^n, is the same as the number of switch states. Therefore if there is a state cannot be reached, there must be two distinct sets of switches (starting from "all off") that produce the same light state. If these two sets are amalgamated, those in both (being applied twice) have no effect. The set remaining when those in both are removed – call it R – leaves the 'all off" state unchanged when its switches are activated, in other words, a choice of switches in which each light is toggled twice. Label the lights and switches from 1 to n so that. say, (switch/light) 2 is in R. Since light 2 is off, it must be toggled by just one adjacent switch, say by switch 3; thus 3 is in R and 1 is not in R. Since 3 is already toggled by itself and switch 2, it is not toggled by 4 and so 4 is not in R. Therefore 5 must toggle 4 and must be in R. Continuing this argument clockwise round the loop shows that R contains a sequence of neighbouring triples consisting of two in R followed by one not in R. The final sequence must eventually tie in with the light/switch 1 which is not in R. Hence the circuit is a amalgamation of these triples and so n is a multiple of three. Therefore if n is not a multiple of 3, all light states can be reached.

Now suppose that n is divisible by 3, say n = 3m. Since activating every third switch turns all the lights on from the "all off" state and since there are three ways of choosing every third switch, we certainly have two distinct switching sets with the same lighting outcome; then by the above argument there must be lighting states that are unobtainable. In particular single lights are unattainable because if they were, every state (which is a combination of suitable single lights) would be attainable. Note that m switches can turn "all off" to "all on" in this case, whereas n are required when n is not a multiple of 3 (by uniqueness and the fact that activating all switches do it).

The lighting circuits in the villa must be chosen from the following lengths (with minimal number switches to go from "all off" to "all on" shown in brackets):
14(14), 3&11(12), 4&10(14), 5&9(8), 6&8(10), 3&4&7(12), 3&5&6(8)

We can rule out 14 because there are at least 2 circuits on each floor. 4&10 is the only option with no circuits of length a multiple of 3 so that must be on the floor where just a single light can be switched on. Since 4&10 needs 14 switches to go from "all off" to "all on", the other two must need a total of 30 - 14 = 16 switches and by inspecting the numbers in brackets we obtain the unique solution 4&10, 5&9 and 3&5&6 on the three floors. There are 7 circuits, of which the longest is 10.

Answer 2304m

If radius of the larger circle is r, then Pythagoras gives OY^2 + r^2 = [OA + r]^2, so OY^2 = OA[OA + 2r], so OY^2 = OA.OB.

OA is a square, hence OB is a square and, since OB = OA + AB, √OA, √AB, √OB give a Pythagorean triple. For example: √OA=3, √AB=4, √OB=5, OY = √OA.√OB = 15

So √AB = √3969 = 63, OA + 63^2 = OB, with √OA = 1, 2, 3 ....22.
Testing gives the only whole number solution as √OA = 16, √OB = 65.
Hence OY = 1040m.

Correspondingly for the smaller circle we have 1040 = OA.[OA + AB].
Since AB is smaller in the smaller circle, √OA must be larger than 16 in this small circle.
For this smaller circle OA + AB = OB and √OA√OB = 1040.
So OA + AB = [1040^2]/OA, so AB = [1040^2]/OA – OA.
Testing √OA = 17, 18 ….…..21, 22 gives √OA = 20, √AB = 48 as the only solution.
So original bridge AB = 48^2 = 2304m.

Answer: 666

Boxes can be made with polygon bases of 3 to 9 sides.

Number of star points=SP=polygon sides=N
Number of faces of box=F=base+folded star points=1+N
Number of edges of box=E=base edges+edges from base vertices to apex=2N

Tabulate [single-figure sum of digits]x[SP+F+E]

Star points	Box faces	Box edges	SP+ F+E	1	2	3	4	Digit sum 5	6	7	8	9
3	4	6	13	13	26	39	52	65	78	91	104	**117**
4	5	8	17	17	34	51	68	85	102	119	136	**153**
5	6	10	21	21	42	63	84	105	126	147	168	189
6	7	12	25	25	50	75	100	125	**150**	175	200	**225**
7	8	14	29	29	58	87	116	145	174	203	232	**261**
8	9	16	33	33	66	99	132	165	198	231	264	297
9	10	18	37	37	74	**111**	148	185	**222**	259	296	**333**

Eight digit sums **in bold** match, but

[117, 153, 225, 261] digit sums divide exactly into none of corresponding SP, F and E values
[150, 111, 333] digit sums divide exactly into two of corresponding SP, F and E values
[222] digit sum divides exactly into just one corresponding SP, F and E value

So knowing that just one of SP, F and E is divisible by the digit sum allows us to know there are 222 sweets per box.

Total sweets = 3 boxes x 222 sweets/box = 666 sweets

Answer: 1301, 1303, 1361, 1367 & 1373

The way that conker values combine means that if you start with a set of conkers and continue a competition until only one is left, then its value at the end will be the sum of all the starting values plus one less than the number of conkers.

There are 11 prime numbers between 1300 & 1400, but, if you simply try all ways of choosing 5 of them, the puzzle will be a hard nut to crack. (There are 462 ways.)

The key to speeding up the calculations is the point that all prime numbers above 3 are either one less or one more than a multiple of 6, because numbers not of those forms are obviously divisible by 2 or 3 (or both).

Carrying on from there, and using the rule for combining values, notice that if a conker with a value of $6a-1$ meets one with a value $6b+1$, then the victor will get a value of $6a-1 + 6b+1 + 1 = 6(a+b) + 1$; and if $6a-1$ meets $6b-1$ the victor will get a value of $6(a+b)-1$; and if $6a+1$ meets $6b+1$ the victor will get a value of $6(a+b)+3$, which cannot be a prime. Examining these consequences shows that the number of 'plus one' conkers stays the same throughout the competition, and that two 'plus one' conkers can never meet. Since the value of the eventually surviving conker, 6709, is one more than a multiple of 6, the competition must have started with exactly one 'plus one' value, and therefore four 'minus one' values.

It is not too difficult to work out 'by hand' all the primes between 1300 & 1400, because it just involves testing numbers for divisibility by primes up to the square root of 1400, that is, up to 37. This reveals that the 'minus one' primes are 1301, 1307, 1319, 1361, 1367 & 1373, and the 'plus one' primes are 1303, 1321, 1327, 1381 & 1399. Then it is a matter of choosing 4 of the 6 'minus one' primes (there are 15 ways of doing this), seeing what the 'plus one' prime would need to be so that the straightforward sum of all of them is 6709-4 = 6705, and then checking whether this wished-for number is actually one of the primes. Success occurs for the primes 1301, 1361, 1367, 1373 and 1303, and for no other set.

(The arithmetic of examining all 15 combinations can be slightly simplified by working instead with the 'number of 6s' in the primes – i.e. the a & b values in the expressions above. This gives for the 'minus one' primes 217, 218, 220, 227, 228 & 229, and for the 'plus one' primes 217, 220, 221, 230 & 233, and for the eventual surviving prime 1118. Then adding 4 of the 'minus one' values to 1 of the 'plus one' values needs to give exactly the 'surviving' value.)

Finally, it should be checked that there is at least one way that the encounters could have happened so that all the intermediate values were prime numbers. One way for this is as follows: 1301 & 1361 met, producing a victor of value 2663, which then met 1303, resulting in a 3967-er victor; 1367 & 1373 met, producing a 2741-er victor; and the 3967-er and 2741-er met in the final encounter, resulting in my 6709-er.

98 PRIME ADVENT CALENDAR

Answer: 73, 79, 83 and 101

In a magic rectangle, all rows add to the same number R, and all columns add to the same number C. Since all the six numbers in a row are prime, they must all be odd (it won't work with one prime being 2), so R must be even. The total of all the numbers is 4R, so it must be divisible by 8.

Similarly, the grand total is 6C, so it is divisible by 3.

Therefore the grand total is divisible by 24.

The minimum value for the grand total is 3+5+7+...+89+97 (i.e. the lowest 24 primes starting with 3). This is 1058. However, the grand total must be divisible by 24, and the next such value is 1080, which is the smallest possible grand total.

We know that the largest prime is 107, and 3+5+...+97+101+103+107 = 1369. For the grand total to be 1080, the three missing primes must total 289. The smallest missing prime must be at least 85 (289-103-101). The candidates are therefore 89, 97, 101 and 103, and the only combination that works is 89+97+103.

The largest primes on the calendar are therefore 73, 79, 83, 101 and 107, the first four of these being opened on the 20th, 21st, 22nd and 23rd.

For completeness, a possible magic rectangle using these numbers is as follows:

29	71	37	83	43	7
101	73	13	31	11	41
47	19	23	61	67	53
3	17	107	5	59	79

[If there was no possible magic rectangle with a total of 1080, then the total would have to be 1104, 1128 etc and there are several possible choices of primes giving these totals, so there could be no unique answer to the Teaser]

Answer: 80

Working on Martha's five, assuming that her single digit was a and that the common difference was b, her tickets would be a, a + b, a + 2b, a + 3b, and a + 4b totalling 5a + 10b or 5(a + 2b). Thus her total must be divisible by 5, i.e. ending in 0 or 5. Thus George's total must display that property and if again we take his single digit as a and common ratio as b, we have his numbers as a ab ab^2 ab^3 ab^4 to total a(1+ b + b^2 + b^3 + b^4) = a(b^5 − 1)/(b − 1)

We now consider values of b and look at the least significant digit of (b^5 − 1)/(b − 1)

b	2	3
(b^5 − 1)/(b − 1)	1 or 6	1 or 6 and so on for the others

Thus, it is clear that a must be 5 to have any chance at all.

We now tabulate possible candidates for b and the results:

2:	5	10	20	40	80	total 155
3:	5	15	45	135	405	total 605
4:	5	20	80	320	too much	

So, George drew one of the top two lines above and we must now solve for Martha's candidates:

5(a + 2b) = 155 giving a + 2b = 31 so that a must be odd and not 5 because George has already got it the following are candidates:

a	1	3	7	9
b	15	14	12	11

So that her numbers would be:

1	16	31	46	61
3	17	31	48	62
7	19	31	43	55
9	20	31	42	53

The last line is ruled out because George has already got 20.
If 5(a + 2b) = 605 then a + 2b = 121 and with a again forced to be odd and not 5, the following are candidates:

a	1	3	7	9
b	60	59	57	56

giving the following as number candidates:

1	61	121	181	241
3	62	121	180	239
7	64	121	178	235
9	65	121	177	233

The only combination featuring digits which add to a perfect square is:

George	5	10	20	40	80
Martha	1	16	31	46	61 giving 49

So, the highest ticket was 80.

Answer: 2,352 square feet

In any rectangle comprising of 'b' rows of 'a' stones the number of stones in the outermost band is
2a + 2b - 4.
In the outermost two bands there are (2a + 2b -4) + 2{(a-2) + 2(b-2) -4} = 4a + 4b - 16 stones. Similarly – 3 bands 6a + 6b -36; 4 bands 8a + 8b - 64; 5 bands 10a + 10b - 100; 6 bands 12a + 12b - 144

As the number of white stones equals the number of outermost red stones then the total number of stones must be twice the number of red stones, so

For 1 red band :	4a+4b-8 = ab :	ab-4a-4b+8 = 0 :	(a-4)(b-4) = 8
For 2 red bands :	8a+8b-32 = ab :	ab-8a-8b+32 = 0 :	(a-8)(b-8) =32
For 3 red bands :	12a+12b-72 = ab :	ab-12a-12b+72 = 0 :	(a-12)(b-12) = 72
For 4 red bands :	16a+16b-128 = ab :	ab-16a-16b+128 = 0 :	(a-16)(b-16) = 128
For 5 red bands :	20a+20b-200 = ab :	ab-20a-20b+200 = 0 :	(a-20)(b-20) = 200
For 6 red bands :	24a+24b-288 = ab :	ab-24a-24b+288 = 0 :	(a-24)(b-24) = 288

The following table gives the factors for the numbers shown in the last column of the above table, the bracketed amounts must equal these factors. As the width is less than 25 feet, bands of 6 and above reds can be ignored as the minimum a/b can be is 24+1. Ignoring the reversibility of a and b, so as not to repeat these factors, this gives values of a and b of –

1 red band			2 red bands			3 red bands			4 red bands			5 red bands		
8	(+4)		32	(+8)		72	(+12)		128	(+16)		200	(+20)	
	a	b		a	b		a	b		a	b		a	b
8x1	12	5	32x1	40	9	72x1	84	13	128x1	144	17	200x1	220	21
4x2	8	6	16x2	24	10	36x2	48	14	64x2	80	18	100x2	120	22
			8x4	16	12	18x4	30	16	32x4	48	20	50x4	70	24
						9x8	21	20	16x8	32	24	25x8	45	28
						12x6	24	18				5x40	25	60
						24x3	36	15				20x10	40	30

Valid solutions are those for which a number for a/b appear in three blocks in the above table (ie they have the same width). Because of the reversibility of a and b the number can appear in either column.

The only number to appear in three columns is 24 giving the following possible solutions –

No. red bands	2	3	4	5	All Combinations of Three 24 ft Rectangles											
					2	3	4	2	3	5	2	4	5	3	4	5
Width	24	24	24	24	24	24	24	24	24	24	24	24	24	24	24	24
Length	10	18	32	70	10	18	32	10	18	70	10	32	70	18	32	70
Total length ft						60			98			112			120	
Total area ft^2						1,440			2,352			2,688			2,880	
Number red stones						720			1,176			1,344			1,440	

The only triangular number of red stones is 1,176 so the total area is 2,352 square feet.

QUICK ANSWERS

Quick Answers

1. 37 and 58571
2. 11-2 and £22
3. 28 cards
4. BvA 1-2 BvC 2-2 BvD 1-0
5. 4, 8, 5, 9, 12, 11, 1, 6, 2, 7, 10, 3
6. George, Kate and Larry
7. 17
8. 16384
9. 143869275
10. (a) 2015 (b) 523
11. 23 and 33603
12. 8778
13. 3/140
14. 106496
15. 20
16. 3A and 9A and 8
17. 32768
18. £249
19. 11 acres
20. 1449
21. (a) HERN and RUSSELL (b) AVERY and DE HOYOS
22. £2-16
23. 6 March
24. £7.84
25. 5, 6 and 9
26. 1, 4, 9, 25, 36, 49, 64
27. 56 and 80
28. 16% (or 1.16 times) faster
29. 9744
30. 99
31. 545/252
32. 26198073

33.	7, 10, 13, 15, 21 and 26
34.	19:56:48
35.	278 and 417
36.	99
37.	3754
38.	54 ounces
39.	8563
40.	243000 sq metres
41.	(a) 123 and (b) 8
42.	2, 5, 8, 10, 25, 40, 50, 100, 250, 500g
43.	10080
44.	£350
45.	147
46.	130
47.	0 1 8 6 7 2 9 5 3 4 0
48.	106:01
49.	(a) 72 and (b) 16
50.	6
51.	15 and 352800
52.	6, 10, 384 and 640
53.	23 125 137 144 and 196
54.	113 cm
55.	2, 3, 5, 9, 17 and 32 grams
56.	289
57.	1854
58.	9
59.	6, 15, 20, 34
60.	20
61.	355
62.	12X523
63.	3 in.
64.	32 cm
65.	1903, 1936, 1969
66.	6 and 7

67.	37mm
68.	2, 3, 41, 58, 69, 70
69.	5520
70.	5832
71.	0, 2, 4, 4, 3, 1, 1
72.	(a) 5 (b) 6 and 8
73.	211
74.	84588800688
75.	£1,764
76.	2385
77.	3 quid and 49 quid
78.	13, 17, 29 and 31
79.	2,601
80.	4177123
81.	'4'
82.	11 hectares
83.	65
84.	19/05/61
85.	1 2 6 8 2 5
86.	7250 seconds
87.	1785
88.	10
89.	4
90.	29
91.	13
92.	530,864,197
93.	56
94.	(a) 7 (b) 10
95.	2304m
96.	666
97.	1301, 1303, 1361, 1367 & 1373
98.	73, 79, 83 and 101
99.	80
100.	2,352 square feet

NOTES

NOTES

NOTES

NOTES

NOTES

NOTES

NOTES